Scott Foresman - Addison Wesley

ENVIRONMENTAL SCIENCE
●●●●●●●●●●●●●●●●●●●●●●●●●●●●●●

LABORATORY MANUAL
TEACHER'S EDITION

Scott Foresman
AddisonWesley

Editorial Offices: Menlo Park, California • Glenview, Illinois • New York, New York
Sales Offices: Reading, Massachusetts • Atlanta, Georgia • Glenview, Illinois •
Carrollton, Texas • Menlo Park, California

Copyright © Addison Wesley Longman, Inc. All rights reserved. No part of this publication may be reproduced, stored in a retrieval system, or transmitted, in any form or by any means, electronic, mechanical, photocopying, recording, or otherwise, without the written permission of the publisher.

Printed in the United States of America

ISBN 0-13-069903-9

1 2 3 4 5 6 7 8 9 10 - ML - 05 04 03 02

Scott Foresman - Addison Wesley

ENVIRONMENTAL SCIENCE
•••••••••••••••••••••••••••••••

LABORATORY MANUAL
TEACHER'S EDITION

CONTENTS

Using the Laboratory Manual	T-7
Cooperative Learning in Science	T-8
Safety and Disposal	T-10
Equipment and Materials List	T-12
Suppliers and Addresses	T-14
Laboratory Annotations	T-15
Laboratory Investigations with Answers	1-144

USING THE LABORATORY MANUAL

Addison-Wesley Environmental Science labs are designed to simulate real research projects and the methods of scientific inquiry. Each investigation familiarizes students with the structure of standard laboratory procedures. Students are encouraged to develop hypotheses, identify and isolate variables, and draw conclusions before, during, and after the activity.

USING THE SE

There is at least one laboratory investigation per chapter of the student text. Each investigation in this manual is divided into a number of sections to allow students to progress in an organized way and to derive and apply useful information.

Introduction: This section provides some general background information as needed. It is mainly intended, however, to motivate students and reveal to them the relevance of the investigation. It also helps to prompt the students to develop a hypothesis on the topic that is being studied.

Goals: Several process-skill objectives are presented to help students grasp the various goals that they should be trying to achieve in the course of the lab investigation.

Lab Warmup: The "Concepts" part of this section activates the student's prior knowledge and correlates the lab to their text. The "Review" part of this section correlates the lab to the appropriate section of the text and lists relevant vocabulary terms relating to principles, procedures, and equipment.

Materials: This section is a list of the equipment, substances, and specimens to be used in the investigation.

Procedure: All procedure steps are spelled out completely. Students will simply be confirming known results or they will be generating variable results from a set procedure.

Laboratory Notebook: Laboratory Notebook pages in each lab simulate the structure of research lab notebooks, indicating predictions, observations, and data. Students enter qualitative and quantitative information and then process that information.

Data Analysis and Conclusions: Students respond to analytical questions that allow them to process and organize data in terms of the original goals of the lab. Questions will encourage them to apply their analysis to their predictions, and to come to appropriate conclusions.

Extension: Finally, students are provided with related discussion topics or with suggestions for further research or investigation.

USING THE TE MANUAL

Front matter of the teacher's edition provides information on cooperative learning, handling live animals in the classroom, safety and disposal procedures, and a list of materials needed for the school year.

Also included are half-page guides for each lab. These guides indicate the time required to complete the lab and the materials needed for a class of 30. Advance preparation for each lab allows you to plan ahead for the specific needs of your class. Trouble-shooting tips, alternative approaches and materials, and teaching tips are also included here.

BEFORE YOUR FIRST LAB

• Explain to the students how a research report is structured, including an introduction, materials list, procedure, and data and conclusions, just like the labs they are about to do.
• Review basic laboratory safety rules contained in the student edition. Make sure the students are familiar with the use and location of safety equipment in the class.
• Use the equipment identification pages of the student edition to review your classroom supplies with the students. Be sure to point out any equipment you will be using that looks substantially different from the illustrations provided.
• Based on the needs and background of your students, you may want to review certain skills, such as graphing and measuring before you begin your first lab.

COOPERATIVE LEARNING IN SCIENCE

Cooperative learning is an approach to teaching that involves building a cooperative climate in the classroom as well as structuring specific group activities. To accomplish a cooperative task, students work in small learning groups in which individuals are assigned specific roles and functions. Often, students find that in sharing information and responsibilities with team members, they come to a better understanding of the science concepts that they are studying. Process and critical-thinking skills are extended as students become aware of the methods teammates use to solve a problem.

Cooperative learning in the science classroom works at two levels:
- Cooperative-learning groups help to enhance the learning of individual members of the group.
- Cooperative-learning groups help to promote collaborative learning and social interaction skills.

COOPERATIVE-LEARNING GROUPS

To establish cooperative-learning groups, the following three steps are recommended:
- Establish academic objectives.
- Determine the size of each cooperative-learning group.
- Assign roles within each group that ensure interdependence.

Cooperative-learning groups should range in number from two to six. The ideal group size, however, is four. Once a cooperative-learning group has been established, the groups should remain together until the assigned activity has been completed.

If a cooperative-learning group is having difficulty working together socially or keeping on task, do not dissolve the group. It is important to keep the group intact because if the group is broken apart, students within the group will not learn the social-interaction skills that are necessary to solve problems or to complete tasks effectively through cooperation and collaboration.

SUGGESTED ROLES IN COOPERATIVE GROUPS

You may establish different-sized cooperative-learning groups based upon the number and kinds of tasks needed to complete an activity successfully.

Despite the fact that specific roles and functions are assigned to individuals, stress to students that each group member is responsible for:
- completing the assignment.
- learning the material.
- making sure that every member of the group has learned the material.
- making sure that every member of the group successfully completes the assignment.

Here are some suggested roles and their jobs that can be assigned with a cooperative-learning group.

Principal Investigator The Principal Investigator is responsible for managing the tasks within the activity, and ensuring that all members understand the goals and content of the activity. The Principal Investigator should read instructions and procedures, check results, and ask questions of the teacher. The Principal Investigator should also act as the facilitator during group discussions.

Materials Manager The Materials Manager is responsible for gathering, assembling, and distributing materials and equipment needed. As an activity progresses, the Materials Manager is responsible for assembling and operating the equipment, as well as checking the results of the activity. The Materials Manager either carries out the investigation or assigns members of the group to carry out the procedure. In addition, the Materials Manager is responsible for ensuring that all equipment is cleaned and returned to its proper place.

Data Collector The Data Collector is responsible for gathering, recording, and organizing the data. The Data Collector also must develop tables, charts, and graphs where needed. Other responsibilities include certifying the data among all group members, and reporting the results of an activity either in writing or orally to the class or to the teacher. If class information is being gathered on a master table on the chalkboard, the Data Collector is responsible for recording the data on the chalkboard.

Timekeeper The Timekeeper is responsible for keeping track of time, for safety, and for monitoring noise level. The Timekeeper also should observe and record the group's social interactions. Other responsibilities are encouraging group members to interact and to discuss the activity as well as to check the results.

COLLABORATIVE/SOCIAL SKILLS

Social skills are basic to the cooperative-learning process. You should assign a specific social skill for each cooperative-learning activity. For example, if the activity requires students to hold a debate, the social skill for the activity can be *listen* carefully. Other cooperative group skills include:
- taking turns.
- sharing resources.
- encouraging participation.
- treating others with respect.
- providing constructive feedback.
- resolving conflict.
- explaining and helping without simply giving answers.
- initiating discussions to solve problems or to make decisions.
- integrating diverse opinions and ideas into a cohesive statement.
- challenging others to seek rationale for answers and conclusions.
- keeping noise levels to a minimum.
- compromising to reach a decision.

SELF-EVALUATION

Encourage students to become actively involved in the evaluation process by providing time for them to reflect on the activity in terms of accomplishments and interactions. Have students examine the success of each activity by discussing the questions listed below.
- What were our tasks/goals?
- Did we accomplish the tasks and goals that we set out to accomplish?
- How well did we meet the social-skills objective?
- How well did we use the process skills that we wanted to use?
- What would we do differently?
- What did we learn from other members of the group?
- What made our group successful? Not successful?
- What could we do to improve communication?
- Rate each other on the ability to compromise, brainstorm and so on (5—high; 1—low).

The *Timekeeper* should take notes and write an evaluation report with the group that is submitted to the teacher.

ESTABLISHING A MATERIALS CENTER

In the classroom or laboratory, set aside an area in which to set up a Materials Center. For each investigation or activity in which students are engaged in cooperative learning groups, place all glassware, equipment, chemicals, and consumable materials in the established Materials Center area. Consistently place out an array of glassware, such as beakers, flasks, graduated cylinders, and so on even if the activity or investigation may not require students to use every piece of glassware. You may also wish to do the same for equipment such as microscopes, balances, spring scales, and so on. By setting out this consistent array, students will be directly involved in decision-making. This will ensure that they decide what equipment is necessary to carry out the investigation. Within the cooperative learning group, initial discussions can center around what materials the *Materials Manager* must assemble for the group to successfully carry out the activity or investigation.

SAFETY AND DISPOSAL

Safety in a laboratory situation is of prime importance in every classroom. The investigations within this biology laboratory manual use materials that are generally safe and nontoxic. However, everyday items such as glassware can be dangerous if not handled carefully. Following safety procedures and guidelines can minimize danger in the laboratory. Good safety habits should be developed and demonstrated by both teacher and students. The following safety guidelines and precautions will help establish safety regulations and procedures in the classroom.

SAFETY SYMBOLS

When appropriate, each investigation provides safety precautions and symbols to alert students to any potential hazards. Because safety in the laboratory is so important, safety precautions appear in boldface throughout the investigation as a reminder to students. An explanation of these symbols appears in the front of the student lab manual.

CLASSROOM ORGANIZATION

All classroom laboratories should be equipped with fully charged fire extinguishers, one or more fire blankets, a first-aid kit, an eyewash station, and a smoke alarm. Students should be aware of the locations of all safety equipment and should know how to use the equipment properly.

The laboratory should be arranged in such a way that equipment and supplies are clearly labeled and easily accessible. Furniture and equipment should be organized with wide, clear aisles to minimize accidents. Supply stations should be set up at various points around the laboratory so students do not crowd one area when gathering supplies. Make sure students keep all work areas clean. Have them remove any unnecessary books, papers, and equipment from lab tables or counters.

PRELAB SAFETY DISCUSSIONS

Begin all laboratory investigations with a discussion of safety procedures. Demonstrate the proper use of materials. Make sure students wear safety goggles during activities that involve potential hazards to eyes. Laboratory aprons also should be worn. Caution students never to taste or eat any substances or chemicals. Review each procedure with students. Caution them to follow the steps in the laboratory investigation in the order the steps are presented.

GLASSWARE

Broken, cracked, or chipped glassware should not be handled with bare hands. Tell students to use heavy gloves and a dustpan for removal and to inform you of any breakage. Provide a container marked "Broken Glass" for disposal. Use only heat-resistant glass in the laboratory. Caution students to use tongs when handling hot glassware.

CHEMICALS

Chemicals should be properly labeled, dated, and stored. Read all caution labels and follow directions carefully. Avoid keeping strong acids in the classroom. When diluting an acid, always add the acid slowly to the water. When using corrosive liquids and/or vapors, use a fume hood. When students are required to note odors and fumes, have them waft the fume towards their nose. They should never inhale fumes directly. Warn students of any chemicals that are flammable.

HEAT AND FLAMES

Have students use hot plates or a water bath rather than a flame whenever possible. Warn them never to allow papers or other flammable substances anywhere near flames. Make sure they tie back hair and loose clothing when near flames, and that they do not use a flame in an open draft. Keep an emergency supply of water and sand nearby to extinguish fires. Remind students that the open end of a test tube that is being heated should always be pointed away from themselves and other students. Be sure all hot plates and burners are turned off when not in use.

STORAGE

Chemicals should be properly labeled and stored in a locked area that is accessible only to authorized persons. Chemicals should be protected from contamination by the outside environment. At the end of an investigation, an inventory should be taken to ensure that all materials have been returned.

DISPOSAL

Always dispose of unused portions of chemicals. Warn students never to return unused chemicals to their original containers. At the end of each investigation, make sure students have disposed of broken glassware and chemicals properly. Provide separate containers for paper, glassware, biological waste, and chemical waste. Students should wash their hands with soap and water when the investigation is completed, and after they have thoroughly cleaned their work stations.

Certain substances should not be poured down the drain, because they are particularly hazardous. Consult the municipal waste treatment authority in your area for details on specific local requirements and regulations. Follow the recommendations scrupulously to avoid damaging the sewage system or posing a threat to the environment.

The disposal of bacterial-culture and fungal culture materials can present a number of problems because of the danger of contamination by potentially deadly pathogens. Reusable equipment, including petri dishes, should be soaked in strong disinfectant and then heat sterilized by autoclaving before being washed. Even materials that are to be disposed of, such as growth media, should be heat sterilized before disposal.

EQUIPMENT AND MATERIALS LIST

The following list includes the equipment and materials you will need for the entire school year. The numbers following each entry indicate the lab numbers for which you will use each material. For specific quantities, refer to the individual lab in the Lab Annos section on page T-17 through T-33.

agar, complete medium   18
alcohol, isopropyl   21
aluminum, pieces of   26
aluminum foil   26
aluminum sheets   23
aphids, turnip   19
apple, pieces of   26
aprons   15
balances   19
beaker clamps   29
beakers,   1-L   26
    250-mL   20, 28
    500-mL   15
    graduated   25
    small   30
beans, dried   6
books of animal skeletons   4
bottles, collection   14
bottles, dropper   13
bowls   4
bread, pieces of   26
Bunsen burners   18, 29
cardboard   8
char extract   11
cheesecloth   4
clocks or watches   25
compost bin, wire or screen   32
computers (optional)   17
containers   30
cotton   25
cotton batting   19
coverslips   13, 14, 28
crude or gear lube oil   15
culture tubes with caps   5
cultures,   *Escherichia coli*, stock   18
cups, clear plastic   27
cups, Styrofoam   25
dishwashing liquid   4, 15
droppers   5, 14, 21, 28, 30
dry ice   21
earthworms   32
Elodea, sprigs of   5
felt, black   21
field guides
    to local plant and animal species   2
    to mammals and birds   4
    to plants   9, 10
    to small animals   10
flags, different colored   3
flashlights   21
flasks, 500-mL   29
forceps, probes or toothpicks   4

glass, pieces of   26
gloves   27
    heat-resistant   29
    protective work   2, 28
glue, fast-drying   21
goggles   15
graduated cylinders,   25
    10-mL   28
    100-mL   20
hammers   22, 23
hot plates   28
jars
    glass, with lids   4, 19
    large, opaque, with tight lids   6, 21
impingers   29
inoculating loops   18
lead wires with alligator clips   24
light meters   3
light source   5
litmus paper   5, 30
magnifying glasses   25
markers   5, 12, 18, 19, 22, 29, 30
matches   18, 29
metric rulers   2, 4, 8, 9, 10, 18, 22, 24, 25, 26
microscopes   28
    compound   13, 14
milliammeters, 0-100mA   24
mold inhibitors,   250-mL   30
motor oil   20
mounting medium   14
nails or tacks   22
nitrogen source   32
notepads   2
nutrient broth   18
orange peel, pieces of   26
paintbrushes, small,   19
paper   7, 8, 16, 26
    graph   7
    thick blotting   21
    white   4
paper bags   7
paper clips   12
paper towels   25, 29, 30
pebbles   20
pellets, barn owl   4
pencils   1, 7
    colored   3, 22, 23, 31
    colored, markers or crayons   1
    grease   11, 28
pens, red or blue felt-tip   6
petri dishes (with covers)   11, 18
pitchforks   32

planting flats   22
plastic, pieces of   26
plastic bags   30
    transparent, with ties   22
plastic bottles with spray pumps   20
plastic buckets, 1-gallon   27
plastic sheeting, opaque   10
plastic tabs   19
pots, 3- or 4-inch   19
protractors   12
pupae, midge   19
radioactive needles inserted into corks   21
rags, thin cloth   21
reference books on hazardous wastes   33
reference materials, library   34
resistors,   20- to 22-ohm   24
right triangles (measuring tool)   2
ringstands   29
rods, glass   26
rubber gloves   15, 20, 21, 22, 25, 26, 30
rubber mallets   2
rubber rings   23
rubber stoppers   23
    2-hole   29
rubber tubing   29
scissors   2, 7, 8, 19, 21, 22
screened cages   19
seeds
    barley, 0 rads   22
    barley, 20,000 rads   22
    barley, 30,000 rads   22
    barley, 40,000 rads   22
    barley, 50,000 rads   22
    buckthorn   11
    buckthorn, heat-treated   11
    cabbage   19
    camas   11
    camas, heat-treated   11
    penstemon   11
    penstemon, heat-treated   11
    whispering bells   11
    whispering bells, heat-treated   11
slides, glass   13, 14, 28
snails   5
soil
    clay   25
    garden   26
    loam   25
    nonsterile   32
    potting   11, 19, 22
    samples of, collected near school   25
    sandy   25
    silt   25
solar cells   24
solutions   0.001% methylene blue   28
    bromothymol blue   5
    neutral red   5
    pH 1   30

    pH 2   30
    pH 3   30
    pH 4   30
    pH 5   30
    pH 6   30
    pH 7   30
stakes   2, 10
    thin wooden,   9
    wooden   12
sticks, wooden popsicle   14
streptomycin   18
string   10, 12, 22
    large balls of   2
    thick   9
tablespoons   15
tacks   23
tape, electrical   23
tape, masking   5, 8, 22
tape measure, flexible   12
test kits
    Dissolved Oxygen   27
    HACH Water Quality   27
    pH   27
    Presence/Absence for Total Coliform   27
test-tube holders   28
test-tube racks   5, 28
test tubes   28
    large   18
thermometers   3, 23, 27, 32
    soil   10
toothpicks   28
trays   21
trowels   9, 10
tubing, glass   23, 29
tubing, plastic   20
twist ties   30
vacuum sources   29
voltmeters (high resistance), 0-1V   24
washers, steel   12
wastes, organic yard and food   32
watches or clocks with second hand   10, 28
water
    containing diatoms   14
    pond, moderately polluted   13
    pond or dechlorinated   5
    pond, slightly polluted   13
    pond, unpolluted   13
    tap   4, 10, 11, 15, 23, 26, 28, 29, 30, 32
    tap, cold   20
    tap, hot   20
water bath   27
watering can or hose   32
wire gauzes   29
wood, pieces of   26
wooden ends, precut   23
yardsticks   12
yeast, dry granules   28

SUPPLIERS AND ADDRESSES

Carolina Biological Supply Co.
2700 York Rd.
Burlington, NC 27215
(800) 334-5551

Central Scientific Co.
3300 CENCO Parkway
Franklin Park, IL 60131
(800) 262-3626

Connecticut Valley Biological
Supply Co., Inc.
P.O. Box 326
Southampton, MA 01073
(800) 628-7748

Edmund Scientific Co.
101 E. Gloucester Pike
Barrington, NJ 08007
(609) 573-6250

Fisher Scientific Educational
Materials Division
4901 W. Le Moyne St.
Chicago, IL 60651
(800) 955-1177

Hubbard Scientific
1120 Halbleib
Chippewa Falls, WI 54729
(800) 289-9299

Lab Safety Supply Inc.
401 S. Wright Rd.
Janesville, WI 53546
(800) 356-0783

Nasco West Inc.
P.O. Box 3837
Modesto, CA 95352
(800) 558-9595

Sargent-Welch Scientific Co.
P.O. Box 5229
Buffalo Grove, IL 60089
(800) 727-4368

Scientific Kit® and Boreal® Laboratories
777 East Park Dr.
Tonawanda, NY 14150
(716) 874-6020

Ward's Natural Science Establishment Inc.
5100 West Henrietta Rd.
P.O. Box 92912
Rochester, NY 14692
(800) 962-2660

LABORATORY ANNOTATIONS

LAB 1

TIME: 30 min

MATERIALS (per class of 30)
pencils
colored pencils, markers or crayons

ADVANCE PREPARATION
none

TEACHING TIPS
- You may wish to review the Concepts section of the lab with students before they begin to graph their data.
- Stress how old Earth is and that its atmospheric changes have occurred over very long periods of time. Point out on students' graphs how close to the present time dinosaurs existed.

LAB 2

TIME: 50 min for sampling
20 min for calculations

MATERIALS (per class of 30)
16 pairs of work gloves
8 metric rulers or tape measures
128 stakes
8 rubber mallets
8 right triangle tools
8 large balls of string
8 pairs of scissors
several guide books to local plant and animal species
8 notepads for recording data

ADVANCE PREPARATION
Plant and animal site
- Choose an area with several different plant and animal species. If you want to simplify the lab, choose an area of relatively uniform appearance. If you wish to make the lab more challenging, assign sites with more diverse plant and animal compositions.
- Identify dangerous plant and animal species beforehand and warn students of these species. You may wish to avoid choosing sites with these species.
- Avoid choosing sites with pronounced abiotic gradients, such as wet/dry or sun/shade.

TEACHING TIPS
- If limited by time or materials, groups can be made larger.
- Students with allergies should wear pollen masks and take other precautions.

LAB 3

TIME: 35 min

MATERIALS (per class of 30)
8 light meters
8 thermometers
48 flags of 6 different colors

ADVANCE PREPARATION

Plant site
Choose an area with an abiotic gradient: some open areas and some shade, or varying moisture, for example.

Flags
- To make the flags, cut strips of construction paper (blue, green, orange, yellow, red, purple) and glue onto popsicle sticks.
- Provide a map of the study site for students to record the location of their flags and surrounding vegetation.

TEACHING TIPS

- Make sure students understand how to make measurements of all three abiotic factors, even though they will measure only one.
- Survey the study site for hazardous plants and animals before beginning the investigation. Students with allergies should wear pollen masks and take other precautions.

LAB 4

TIME: 50 min

MATERIALS (per class of 30)
15 barn owl pellets
15 metric rulers
15 glass jars with lids
water
dishwashing liquid
30 pieces of cheesecloth cut into 6-cm squares
15 bowls
30 sheets of white paper
15 forceps, probes, or toothpicks
field guides to mammals and birds
books of animal skeletons

ADVANCE PREPARATION

Owl pellets
Order the barn owl pellets in advance from a biological supply company, or find your own in wooded areas, parks, or old farm buildings. If collecting your own, dry them first, then fumigate them in polyethylene bags with naphthalene.

Cheesecloth
Cut the cheesecloth into 6-cm squares.

TEACHING TIPS

- Let students know they need not find, identify, and match every small bone, as long as the major bones are identified. The skulls are most important.
- If students find more than five animals in their pellets, they may continue the chart in the Lab Notebook on a separate sheet of paper.
- Make a class data chart on the chalkboard and have students calculate percentages for each species. For example,

$$\% \text{ species A} = \frac{\text{total \# species A}}{\text{total \# of all animals recorded}} \times 100.$$

LAB 5

TIME: 30 min to set up tubes
7 days for observation

MATERIALS (per class of 30)
64 culture tubes with caps
8 markers
masking tape
litmus (pH) paper
pond or dechlorinated water
16 droppers
bromothymol blue solution
neutral red solution
64 *Elodea* sprigs
32 snails
8 test-tube racks
light source

ADVANCE PREPARATION

Supplies
Order supplies in advance from a biological supply company. You may use *Anacharis* instead of *Elodea*.

Indicator Solutions
Bromothymol blue and neutral red solutions may be purchased premixed. To make your own solutions:
Dissolve 0.1g bromothymol blue in 1L distilled water. If the solution does not appear blue, add drops of 4% sodium hydroxide solution until the bromothymol solution turns blue.

Pond Water
If pond water is not available, let tap water stand for 24 hours before using. Check the pH of the pond water. It should be close to neutral. The litmus paper should preferably cover a fairly narrow range, such as 4-10.

TEACHING TIPS

- Bromothymol blue changes to yellow at an approximate pH of 6.0. Neutral red changes to yellow at pH 8.0.
- Make sure the light source does not cause the tubes to get too warm.
- If students get unexpected results, have them try to explain these results. Discuss methodology, errors, and chance.

LAB 6

TIME: 50 min

MATERIALS (per class of 30)
8 large, opaque jars or cans with tight lids
2 bags of dried beans
8 red felt-tip pens
8 blue felt-tip pens

ADVANCE PREPARATION
none

TEACHING TIPS

- Emphasize that students should not look at the beans as they remove the second sample from the container.
- Have students compare the estimates of population size calculated in the two experiments. Discuss possible reasons for differences in the estimates (chance, different sample sizes, experimental errors, etc.).

LAB 7

TIME: 50 min

MATERIALS (per class of 30)
blank paper or directional slips as described in the lab
30 scissors
30 paper bags
30 sheets of graph paper
access to a photocopier

ADVANCE PREPARATION

Directional slips
You may wish to prepare the directional slips ahead of time, although students are instructed to do so. Photocopy the directions listed in the Procedure section and cut them out, placing each set of slips into a separate paper bag. All slips should be the same size.

TEACHING TIPS

- Separate students into groups of 5. Each student will perform each step of the procedure individually.
- Stress the randomness of mutations and natural selection. Explain to students that their selections of traits represent the demands of the organism's habitat.
- You may wish to go over the answers after the Lab Notebooks are graded to dispel any misconceptions students may have about natural selection and evolution. Stress that evolution does not move toward a specific, ideal goal. Evolving populations simply respond to pressures from the environment.

LAB 8

TIME: 50 min

MATERIALS (per class of 30)
10 metric rulers
10 sheets of paper
10 scissors
large piece of cardboard
10 rolls of masking tape

ADVANCE PREPARATION

Mouse and owl squares
To save lab time, you might cut out and provide students with the mouse squares and owl squares.

TEACHING TIPS

- Suggest that students use two different colors when plotting the data for the owl population and the mouse population on their graphs.
- Have groups of students compare their graphs to reinforce the concept that the sizes of predator populations and prey populations change in a cyclic manner. The cycles are also coupled to each other.

LAB 9

TIME: 50 min

MATERIALS (per class of 30)
20 thin wooden stakes
field guides to plants
10 6-m lengths of string
10 trowels
10 metric rulers

ADVANCE PREPARATION
String
Cut string into 6-meter lengths.

TEACHING TIPS
- Assign tasks to each member of a group. For example, one student identifies the plants along the transect, another student compares the plants' roots, while a third student measures the heights of the plants.
- Students may need help in recognizing successional stages within the lot or field they are studying. Emphasize that a community changes over time until it develops into a stable climax community.
- Be sure to identify any hazardous plants and animals in the study site and warn students beforehand. Students with allergies should wear pollen masks and take other precautions.
- Have students wash their hands thoroughly with warm, soapy water after this activity.

LAB 10

TIME: 50 min for observations of microhabitats and to change conditions within them
3 weeks of maintaining changed conditions in microhabitats
20 min for observations of microhabitats after three weeks

MATERIALS (per class of 30)
8 metric rulers
64 stakes
16 5-m lengths of string
8 thermometers
8 trowels
water
8 watches that indicate seconds
field guides to plants
field guides to small animals
opaque plastic sheeting

ADVANCE PREPARATION
String
Cut string into 5-meter lengths.

TEACHING TIPS
- Have a representative from each group return to the microhabitats to take a second recording of soil temperature, either in the morning or in the afternoon as needed.
- Be sure to survey sites beforehand for dangerous plants and animals. Students with allergies should wear pollen masks and take other precautions.
- Instruct students to take the soil temperature at equal depth (several inches) in each site.

LAB 11

TIME: 50 min
10-min observations per week over a period of four weeks

MATERIALS (per class of 30)

160 camas *(Camisonia californica)* seeds, 80 of them heat treated
160 buckthorn *(Ceanothus megacarpus)* seeds, 80 of them heat treated
160 whispering bells *(Emmenanthe penduliflora)* seeds, 80 of them heat treated
160 penstemon *(Penstemon spectabilis)* seeds, 80 of them heat treated
64 petri dishes with covers
8 grease pencils
2 bags fine potting soil
water
4 liters char extract (water in which charred and ground birch wood has been soaked)

ADVANCE PREPARATION

Seeds
You can order seeds of the four required species from the Theodore Payne Foundation, 10459 Tuxford Street, Sun Valley, CA 91352. For approximately six weeks before the experiment is performed, you should store all the seeds in closed containers in a refrigerator, to provide the cold treatment that is important for eventual germination. To heat-treat the seeds, bake them for 30 minutes in an oven set at 80°C a few hours or the day before the experiment.

Charred wood extract
To prepare the charred wood, first burn balsa or birch dowels until their outer surface is blackened, without consuming the inner part of the wood or burning the wood to ash. Soak the charred wood in water for 2 days, and have students use the resulting solution as an extract with which to moisten the soil.

TEACHING TIPS

- Emphasize to students that clear labeling is essential in this experiment, given the number of variables they must be able to distinguish.
- Point out that it is important to control variables, such as the amount of water and amount of light.
- You may wish to review construction of bar graphs and calculation of percentages for students who require assistance with these procedures.

LAB 12

TIME: 15 min for clinometer construction
90 min for on-site measurements, not including travel time

MATERIALS (per class of 30)

10 flexible tape measures
10 paper clips
1 spool of string
10 protractors
10 steel washers
10 yardsticks
40 wooden stakes
10 markers

ADVANCE PREPARATION

Area studied
Search out an appropriate and safe location of typical tree density in your area. Obtain permission, if necessary, for the class to use it.

Construction information
You may wish to provide information on the number of board feet of wood needed to construct a house typical of your area, rather than having students seek out the information themselves. If so, contact a local builder before doing the investigation.

TEACHING TIPS

- Make sure that students exercise caution in their on-site investigation. Suggest that student partners "spot" for each other when they site trees for height estimates. Insist that the class remain within the selected area, and keep students under strict supervision.
- Be sure to identify any hazardous plants and animals in the study site and warn students beforehand. Students with allergies should wear pollen masks and take other precautions.
- Emphasize the theme of resource use and conservation in this investigation.

LAB 13

TIME: 50 min

MATERIALS (per class of 30)

- 50 mL of pond water from an unpolluted site
- 50 mL of pond water from a slightly polluted site
- 50 mL of pond water from a moderately polluted site
- 3 large dropper bottles
- 30 glass slides
- 30 coverslips
- 10 compound microscopes

ADVANCE PREPARATION

Pond water

Obtain samples from various freshwater sources in your area that you know differ significantly in water purity and in plankton diversity, but are similiar in terms of other abiotic factors. You may wish to order the samples from a biological supply house. Shake each of the samples thoroughly just before the investigation to distribute the plankton.

TEACHING TIPS

- Review slide preparation techniques and make sure that students know how to use a compound microscope.
- Because the possible number of kinds of plankton is so large, you may wish to help students with some of the more difficult identifications and explain how they can use features of their microscopes to help distinguish the different types.
- To improve the accuracy of results, you may have students work with larger samples of pond water. If so, allow more time for counting and identification.
- Have students wash their hands with warm, soapy water after this activity.

LAB 14

TIME: 50 min, not including on-site collection time

MATERIALS (per class of 30)

- 6 water samples containing diatoms
- 6 collection bottles
- 6 droppers
- 6 coverslips
- 1 small bottle of mounting medium
- 6 glass slides
- 6 compound microscopes

ADVANCE PREPARATION

Collection sites

Select collection sites that are safe for students to access. The water at the various sites should differ in its level of pollution. Do not select sites that are so dangerously polluted that they represent a hazard to students.

Oven-dried coverslips

Dry the coverslips in an oven at 220°C for 1 to 3 hours, until no black carbon spots are visible. Be careful not to mix up the students' coverslips. You could place the coverslips in the oven systematically, with positions keyed to the names of the students.

TEACHING TIPS

- Review the concepts of diversity and environmental quality, which were covered in Section 4.2, Diversity and Stability.
- Advise students not to make collections until at least one week after heavy rains, otherwise diversity data will not be representative.
- Stress that students make their collections in safe locations and that an adult should accompany them. You may wish to provide a list of acceptable sites.
- Make sure students understand that they are comparing each diatom with the preceding one only. Also be certain they recognize that they are obtaining information on diversity. Explain the concept of the diversity index, and make sure they calculate and apply it correctly.
- Emphasize that students avoid dangerous sites, such as deep-water areas during collection. Explain that they should look for brownish coatings on wet objects at the site and scrape off some coating using a popsickle stick, and place this material into a collection bottle. They should add to the bottle a small amount of water skimmed from the surface.
- Have students wash their hands with warm, soapy water after handling samples.

LAB 15

TIME: 30 min Part A
30 min Part B, 1 week later

MATERIALS (per class of 30)
10 500-mL beakers
1 quart crude oil or gear lube oil
10 tablespoons
1 standard container dishwashing liquid
30 pairs rubber gloves

ADVANCE PREPARATION

Water and oil mixture
In each beaker, mix about 1 tablespoon of oil with about 250 mL of water.

TEACHING TIPS

- Make sure you review students' proposed techniques for oil removal. Do not comment on the probable effectiveness of the techniques, but do prohibit any potentially harmful ones, such as burning or the addition of dangerous chemicals. Encourage creativity.
- Tell students to wear old clothing on the days they perform the investigation.
- In judging the effectiveness of techniques, consider whether the water looks clean of oil, whether the beaker's inside or rim is smeared with oil, whether any absorbent materials used have been removed effectively, etc.

LAB 16

TIME: 50 min

MATERIALS (per class of 30)
15 sheets of paper, labeled and cut into slips

ADVANCE PREPARATION

Paper slips
Prepare a sheet of paper containing a list of at least 25 garbage items of various disposal categories. Such items might include: newsprint, rubber band, eggshells, brown glass bottle, insecticide can, vegetable peelings, cereal box, aluminum can, used cat box litter, spoiled ground beef, bleach bottle, broken air conditioner, colorless glass jar, plastic detergent container, child's slightly worn clothing, coffee grounds, can containing paint, styrofoam cup, etc. Also write down boldly and clearly the following four disposal-option terms: RECYCLE, REUSE, COMPOST, LANDFILL. Make a photocopy of the list for every student group. Carefully cut each photocopy to generate separate slips with one trash item or disposal option on each.

TEACHING TIPS

- Before students carry out the investigation, discuss the general characteristics of trash items that are suitable for disposal by each method. Also discuss the importance of proper disposal from an environmental perspective.
- A week or two after the investigation, follow up to see whether students have in fact altered their behavior with regard to buying and consumption or reuse of products, generating trash, and disposing trash as a result of having performed the investigation.

LAB 17

TIME: 50 min

MATERIALS (per class of 30)

8 computers (optional)

ADVANCE PREPARATION

None

TEACHING TIPS

- The mathematics involved may be very difficult for some students to understand. Review the concepts and the calculation methods thoroughly. Also, demonstrate what is meant by the use of a smooth curve to connect data points, and explain the technique of extrapolation.
- Be sensitive to differing philosophies regarding the issue of limitation of population growth that may arise during class discussion.

LAB 18

TIME:
- 15 min part A
- 30 min to collect data and inoculate gradient dishes
- 15 min part B
- 30 min to collect data

MATERIALS (per class of 30)

48 petri dishes	6 inoculating loops
complete agar medium	6 Bunsen burners
complete nutrient broth	matches
stock culture of *E. coli*	12 large glass test tubes
streptomycin	6 metric rulers
6 markers	

ADVANCE PREPARATION

Supplies

Stocks of *E. coli*, streptomycin, complete broth, and complete agar can be obtained through a biological supplier.

Agar dishes

Mix agar and pour all dishes ahead of time. Make sure you label all the dishes with the concentration of streptomycin. Indicate on each gradient dish where the concentration of streptomycin is zero. The following dishes are needed for this lab:

6 complete media
6 complete + 20µg/mL strep
12 complete + 200µg/mL strep
6 complete + 1000µg/mL strep
6 gradient – 0-20 µg/mL strep
6 gradient – 0-200 µg/mL strep
6 gradient – 0-1000 µg/mL strep

Making gradient dishes

To make gradient dishes, pour about 10 mL of complete agar containing no streptomycin into a petri dish. Tilt the dish and prop it on a glass rod such that the agar barely reaches one end of the dish. Let cool. Mark the dish zero at the point where the agar ends. Directly across the dish, mark it with the maximum concentration of streptomycin. Then mix a batch of agar with the maximum concentration of streptomycin (20, 200, or 1000 µg/mL). Pour into the dish resting on a flat surface. Let cool. The streptomycin will diffuse into the agar beneath it, resulting in a concentration gradient across the dish.

TEACHING TIPS

- Some students may need help in calculating the MIC values.
- If a student group does not observe a mutant colony on any of the gradient dishes, a colony from another group can be used.
- Have students wash their hands with warm, soapy water after handling anything containing *E. coli*.

LAB 19

TIME: 20 min to set up experiment
2 to 3 weeks growth period
50 min to collect and analyze data

MATERIALS (per class of 30)

cabbage seeds (*Brassica oleracea capitata*)
75 3- or 4-inch pots
potting soil
75 plastic tabs for labeling plants
30 markers
3 screened cages
500 turnip aphids (*Hyadaphis pseudobrassicae*)
30 small paintbrushes
250 midge pupae (*Aphidoletes aphidimyza*)
30 scissors
60 jars with lids
balances
cotton batting

ADVANCE PREPARATION

Screened cages
Screened cages may be obtained from a biological supply house, or built from wood and 40 μm mesh polyester screening. Seal any cracks with silicon caulk. Cover all sides of the cage with the polyester screening, including the bottom.

Aphids and midges
Obtain turnip aphids and midges from biological supply companies. Aphids can be also be obtained by picking them off of cabbage seedlings planted outside in the spring. Be sure they are the turnip aphid (clear green) and not the cabbage aphid (grey-white). Maintain them in separate enclosures, with weekly supplements of cabbage seedlings. Maintain midge pupae in a separate enclosure with plenty of aphid-infested cabbage. About 10 to 12 days before the lab, collect leaves with 3rd to 4th instar midge larvae and place these on a tray covered with wet cotton batting.

Cabbage plants
Plant cabbage seeds two or three weeks before the students set up the experiment. Transfer healthy seedlings of approximately equal size to 3- to 4-inch pots (one per pot) to give to the students. If the climate or season is inappropriate for growing cabbage plants, perform the experiment in a greenhouse.

TEACHING TIPS

- If you wish to have each student be responsible for one plant in each cage, increase or decrease the number of plants in each cage accordingly. Otherwise, divide the 75 plants among the students.
- Create a class data table on the chalkboard for students to record their data.

LAB 20

TIME: 50 min

MATERIALS (per class of 30)

30 pairs of rubber gloves
10 plastic bottles with spray pumps
pebbles to fill all the bottles halfway
10 100-mL graduated cylinders
1 L motor oil
10 pieces of plastic tubing to fit pump nozzles
30 250-mL beakers
cold tap water
hot tap water

ADVANCE PREPARATION

Tubing
Cut the plastic tubing into approximately 30-cm lengths to use as external tubing.

TEACHING TIPS

- Check on the setup of the apparatuses before students proceed with the investigation.
- Review the technique of bar graphing for the benefit of students who may have difficulties in this area.
- The motor oil may be collected and reused by the next class. Dispose of the motor oil as required by your community.
- In judging the effectivness of techniques, consider whether the water looks clean of oil, whether the beaker's inside or rim is smeared with oil, whether any absorbent materials that have been added have been removed effectively, etc.
- Use oil with a lower viscosity—10w-30 or lower—for best results. (Lower numbers indicate lower viscosity.)

LAB 21

TIME: 40 min

MATERIALS (per class of 30)
8 rubber gloves
8 scissors
8 black felt squares
8 large glass jars with lids
8 thick blotting paper sheets
fast-drying glue
8 droppers
isopropyl alcohol
dry ice
8 trays
thin cloth rags
8 flashlights
8 radioactive needles inserted into a cork (or other source of alpha radiation)

ADVANCE PREPARATION

Radiation source
To save time, cloud chambers with an alpha radiation source can be purchased through a biological supplier. The radioactive needle (or other source of alpha radiation) can also be purchased separately. Follow directions for using the radiation source carefully.

Dry ice
Obtain dry ice from a fishing bait store or ice cream shop. Place the dry ice in trays before class begins so students will not need to handle it. You may wish to use rags to level the chambers and slow sublimation of the dry ice.

TEACHING TIPS

- It is good practice to have students handle the radiation source carefully, but ensure them that this radiation will not harm them.
- Darken the room and use powerful flashlights to observe the vapor trails.
- Make sure students return the radiation source at the end of the investigation.

LAB 22

TIME: 50 min to set up
20 min each week for three weeks to record data

MATERIALS (per class of 30)
30 rubber gloves
6 planting flats
6 rolls of masking tape
6 markers
3 bags potting soil, moistened
6 hammers
48 nails or tacks
string
6 scissors
packets of barley seeds (6 each of 5 different levels of radiation exposure: 0, 20,000, 30,000, 40,000, and 50,000 rads)
30 metric rulers
6 transparent plastic bags with ties
colored pencils (6 each of 5 colors)

ADVANCE PREPARATION

Seeds
Order irradiated barley seeds through a biological supplier.
Plan to plant the seeds on Monday, and measure them each Monday for three weeks afterward.

TEACHING TIPS

- If mold develops on the soil, take the flat out of the plastic bag to aerate.
- If time allows, have students pool their data and make class graphs.
- Have each student count and measure one row of seedlings (one radiation dosage) for the duration of the experiment, but record all data from the group.

© Copyright Addison-Wesley. All rights reserved.
Addison-Wesley Environmental Science Laboratory Manual

LAB 23

TIME: 20 min to build the concentrators
60 min to record data

MATERIALS (per group of 30)

6 aluminum sheets
12 precut wooden ends
tacks
6 hammers
6 pieces of glass tubing
6 rubber stoppers
electrical tape
water
6 thermometers
6 rubber rings to fit around thermometers
colored pencils

ADVANCE PREPARATION

Wooden Ends
Make parabolic wooden ends as shown in Figure 23.3 in the student lab book.
Draw the pattern on the wood and cut it out. Drill a hole a little larger than the diameter of the glass tubing you will use at the focal point of each parabola. You may wish to have a shop class make these. Use a math book as a reference to determine the exact shape of the end pieces.

Aluminum Sheets
Cut six sheets of shiny aluminum 25 cm long, with a width to fit around the curved portion of the wooden ends.

Glass Tubing
Select a diameter of glass tubing small enough to fit inside the drilled holes, but larger than the diameter of the thermometer. Cut it into six 30-cm lengths.

TEACHING TIPS

- Explain the concept of the parabola to students before the lab.
- For best results, choose a time and season such that the sun is not positioned directly overhead.
- If time allows, have students pool their data. The results should be similar.

LAB 24

TIME: 50 min for Part A
50 min for Part B

MATERIALS (per class of 30)

12 lead wires with alligator clips
6 voltmeters (high resistance), 0-1V
6 milliammeters, 0-100mA
6 20- to 22- ohm resistors
6 solar cells
metric rulers

ADVANCE PREPARATION

Circuits
You may wish to set up one test circuit for the entire class and have groups take turns testing their solar cells.

Solar Cells
Test the solar cells before the lab to be sure that they register on the voltmeters and ammeters.

TEACHING TIPS

- You may wish to provide a sample calculation of the efficiency of the solar cell for students who are having difficulty with the math.

LAB 25

TIME: 50 min

MATERIALS (per class of 30)

30 pairs of rubber gloves
1 bag sandy soil
1 bag silt soil
1 bag clay soil
1 bag loam soil
15 magnifying glasses
soil samples collected near school
15 styrofoam cups
2 rolls paper towels
cotton
15 metric rulers
15 graduated cylinders
15 beakers with graduations
clock or 8 watches

ADVANCE PREPARATION

Soil samples
Collect soil samples of clay, sand, silt and loam using the criteria found in the lab. Allow the students to practice identifying the different soil types before they try to identify their own samples.

Soil samples to be collected by students
Examine sites around the school in advance. Find several that vary substantially in soil composition. Assign one area to each group of students. If soil collection by students is impractical, you may wish to collect all the samples yourself.

TEACHING TIPS

- Supervise students carefully. Make sure that they collect samples only from the assigned areas.
- If graduated beakers are not available in sufficient numbers, you may wish to have students calibrate unmarked beakers themselves.
- Stress the importance of using techniques consistently from group to group to assure reliability of data.

LAB 26

TIME: 30 min to set up investigation
15 min to record observations each week for 4 weeks

MATERIALS (per class of 30)

30 pairs of rubber gloves
15 1-L beakers
2 bags garden soil
15 pieces of each of the following materials: glass, paper, aluminum, orange peel, apple, bread, wood, and plastic
15 metric rulers
water
1 small roll aluminum foil
15 glass rods

ADVANCE PREPARATION

Model Landfill Objects
Make the objects similar in size and shape. Thin squares about 2 cm wide would be suitable. Make sure that none of the objects, including the glass, is sharp.

TEACHING TIPS

- Make sure that students place the objects so that they are readily visible without disturbing the soil.
- Encourage students to make careful observations that are as quantitative as possible.
- Use the results of the investigation as the basis for a class discussion.

LAB 27

TIME: 50 min

MATERIALS (per class of 30)

30 pairs of gloves
8 1-gallon plastic buckets
8 clear plastic cups
8 HACH Water Quality Test Kits
pH Test Kit
Dissolved Oxygen Test Kit
Presence/Absence Kit for Total Coliform
8 thermometers
water bath

ADVANCE PREPARATION

Test kits
Order test kits from HACH Company, 5600 Lindbergh Dr., Loveland, CO 80539. Similar test kits may be available through other suppliers, but the procedures may vary.

Water samples
These tests were designed for freshwater sources. You may wish to obtain water samples before class begins.

Water bath
Heat the water bath at the beginning of the session.

TEACHING TIPS

- You may wish each group to test a different source of water, such as tap water, stream water, or pond water.
- In the dissolved oxygen test, the settling of powder on the bottom of the bottle will not affect test results.
- Emphasize that students should avoid trapping air bubbles in the dissolved oxygen test and introducing contaminants in the coliform test.

LAB 28

TIME: 50 min

MATERIALS (per class of 30)

42 test tubes
6 grease pencils
6 10-mL graduated cylinders
water
6 test-tube racks
72 dry yeast granules
30 pairs of protective gloves
6 250-mL beakers
6 hot plates
6 test-tube holders
6 watches or a clock that indicates seconds
6 droppers
6 glass slides
100 mL 0.001% methylene blue solution
6 toothpicks
6 coverslips
6 microscopes

ADVANCE PREPARATION

Methylene blue
Add 0.1 g methylene blue to enough water to make 100 mL of solution.

TEACHING TIPS

- Make sure students observe safety guidelines to avoid burns.
- Stress the importance of using the same quantities of water and yeast in all the test tubes.
- Visit work stations to be certain that students can recognize the difference between live and dead yeast cells. The color difference should be sufficient to make differentiation possible but will not be obvious in all cases.

LAB 29

TIME: 50 min

MATERIALS (per class of 30)

24 2-hole rubber stoppers
24 5-cm pieces of glass tubing
30 impingers (15-cm pieces of glass tubing with a smaller opening at one end)
8 15-cm pieces of glass tubing
8 paper towels
24 500-mL flasks
water
24 40-cm lengths of rubber tubing
8 ringstands
8 beaker clamps
8 wire gauzes
8 Bunsen burners
8 markers
8 vacuum sources
30 pairs of heat-resistant gloves
matches

ADVANCE PREPARATION

Impingers
Make the impingers by heating and drawing out one end of each piece of glass tubing to produce a narrow opening.

Paper towels
Select paper toweling that contains no special coatings, colorings, or inks.

TEACHING TIPS

- Review safety guidelines before carrying out the investigation.
- Check the apparatuses for safety and airtightness before allowing students to ignite the burners.
- You may wish to have students carry out the experiment in a fume hood. In any case, make sure that the room is well ventilated.

LAB 30

TIME: 50 min to set up investigation
15 min to record data on each of two later days

MATERIALS (per class of 30)

30 pairs of rubber gloves
375 seeds
15 small beakers
250 mL mold inhibitor
1 liter each of solutions pH 1 through 7
15 containers to hold solutions
litmus paper
water
2 rolls paper towels
15 plastic bags
15 droppers
15 twist ties
15 markers

ADVANCE PREPARATION

Acid solutions
Start with 1-molar (1-M) sulfuric acid to make the assigned solutions. Successive dilution ratios of acid to water should be approximately as follows: pH 1, 1 mL of 1-M sulfuric acid for every 10 mL of solution desired; pH 2, 1 mL of pH-1 acid just produced for every 10 mL of solution desired; pH 3, 1 mL of pH-2 acid for every 10 mL desired; pH 4, 1 mL of pH-3 acid for every 10 mL desired; pH 5, 1 mL of pH-4 acid for every 10 mL desired; pH 6, 1 mL of pH-5 acid for every 10 mL desired; for pH 7, use pure water without any acid.

TEACHING TIPS

- Use only fast-sprouting seeds, such as beans.
- Check accuracy of students' pH readings of solution before having them proceed.
- Some students may need help with averaging and graphing the class data.

LAB 31

TIME: 50 min

MATERIALS (per class of 30)

30 colored pencils

ADVANCE PREPARATION

None required

TEACHING TIPS

- Review with students abiotic and biotic factors that affect the survival of organisms.
- Point out to students that although the island-reserve scenario is a simplified and artificial one, the principles involved do apply to actual situations.
- The determination of the final habitats may be difficult conceptually for many students who must weigh simultaneously the combined effects of many factors on migrating organisms. Remind students that, in deciding whether a species will survive, they must think in terms of several factors: the abiotic conditions that support the life of the organisms, the final location (and existence) of food sources, and the rates at which the organisms can migrate to reach habitats with suitable abiotic conditions and food supply.

LAB 32

TIME: 40 min to set up compost pile
10 min each day for 21 days to record data

MATERIALS (per class of 30)

wire or screen compost bin
organic yard and food wastes
dirt (nonsterile soil)
nitrogen source (optional)
water
watering can or hose
10 to 20 earthworms
thermometer
pitchfork

ADVANCE PREPARATION

Organic wastes
Tell students before the lab to bring kitchen scraps from home. Make arrangements with the school maintenance crew to save grass clippings and raked leaves.

TEACHING TIPS

- Extend the lab to 28 days if necessary.
- Put the compost to good use in school gardens or landscapes.
- You may wish to start a compost program for your school, using yard waste and scraps from the cafeteria. Use a 3-bin system: one for adding waste, one for the decomposition process, and one for ready-to-use compost. Be sure everyone involved is well informed about what can and cannot be added to the pile.
- Have students wash their hands with warm, soapy water after handling earthworms, compost materials, and/or dirt.

LAB 33

TIME: 50 min

MATERIALS (per class of 30)
reference books on hazardous wastes

ADVANCE PREPARATION

Surveys
Consult with faculty and staff members in advance to arrange a time for students to conduct their surveys. Be sure to warn students about any hazards in the areas they survey.

TEACHING TIPS

- Warn students not to handle hazardous chemicals. Have them report any spills or other problems to you.
- Make sure a teacher or other responsible adult is with the students at all times.
- Stress that it is expensive for industries and small-quantity generators to handle hazardous wastes according to government guidelines, and that usually the consumer pays indirectly for the handling of hazardous wastes by paying more for products.

LAB 34

TIME: 50 min for groups to prepare their arguments after researching the issues
50 min for role-playing and discussion

MATERIALS (per class of 30)
library reference materials

ADVANCE PREPARATION

Interest groups
Divide the class into six groups: chemical company, environmental agency, residents for restrictions, residents against restrictions, city council, and court.

TEACHING TIPS

- After assigning groups and explaining the lab, give students ample time to research the issues. Meet again about a week later.
- Be as objective as possible, and encourage students to be open to other people's opinions.

Scott Foresman - Addison Wesley

ENVIRONMENTAL SCIENCE
• •

LABORATORY MANUAL
TEACHER'S EDITION

Scott Foresman
Addison Wesley

Editorial Offices: Menlo Park, California • Glenview, Illinois • New York, New York
Sales Offices: Reading, Massachusetts • Atlanta, Georgia • Glenview, Illinois •
Carrollton, Texas • Menlo Park, California

Copyright © Addison Wesley Longman, Inc. All rights reserved. No part of this publication may be reproduced, stored in a retrieval system, or transmitted, in any form or by any means, electronic, mechanical, photocopying, recording, or otherwise, without the written permission of the publisher.

Printed in the United States of America

ISBN 0-13-069903-9

1 2 3 4 5 6 7 8 9 10 - ML - 05 04 03 02

CONTENTS

Safety Guidelines		5
Doing Science		7
Lab 1	The Atmosphere and Living Things	9
Lab 2	Sampling a Biotic Community	13
Lab 3	Abiotic Factors	17
Lab 4	Owl Pellets	21
Lab 5	The Oxygen Cycle	25
Lab 6	Estimating Population Size	29
Lab 7	Modeling Evolution	33
Lab 8	Predator/Prey Interaction	37
Lab 9	Ecological Succession	41
Lab 10	Schoolyard Microenvironments	45
Lab 11	Chaparral and Fire Ecology	49
Lab 12	Forestry and Conservation Study	53
Lab 13	A Survey of Plankton Communities	57
Lab 14	Diatoms as Water-Quality Indicators	61
Lab 15	Cleaning Up Oil Spills	65
Lab 16	Garbage Disposal	69
Lab 17	Human Population Growth	73
Lab 18	Detecting Mutant Bacteria	77
Lab 19	Biological Pest Control	81
Lab 20	Oil Extraction	85
Lab 21	Observing Radiation	89
Lab 22	The Effects of Radiation on Plants	93
Lab 23	Solar Energy Concentrator	97
Lab 24	Electricity from Solar Cells	101

Lab 25	Soil Exploration	105
Lab 26	Landfill Biodegradation	109
Lab 27	Testing Water Quality	113
Lab 28	Thermal Pollution	117
Lab 29	Modeling a Wet Scrubber	121
Lab 30	Acid Rain and Seed Growth	125
Lab 31	Global Warming and Biodiversity	129
Lab 32	Composting	133
Lab 33	Hazardous Wastes Survey	137
Lab 34	Environmental Issues and Public Policy	141

SAFETY GUIDELINES AND SYMBOLS

An environmental science laboratory is a place where many exciting things can happen. It is also a place with dangerous materials and potential hazards. Following sensible safety precautions will help to ensure that your experience in the lab is a positive one. Read the following guidelines before you begin working in the laboratory, and review them from time to time throughout your study of environmental science.

1. Read through the procedures of each laboratory activity before you come to class, so that you are familiar with them.
2. Know how to locate and use all safety equipment in the laboratory, including the fume hood, emergency shower, first aid kit, fire blanket, fire extinguisher, and eyewash. Also be sure to locate the nearest exit in case of an emergency.
3. There must be no horseplay, running, or other behavior that can be dangerous in the laboratory.
4. Always conduct your experiments with someone else present and under adult supervision.
5. Wear safety goggles when handling all hazardous chemicals, working with an open flame, or when otherwise instructed.
6. Wear an apron or a smock to protect your clothing in the laboratory.
7. Tie back long hair, and secure any loose-fitting clothing.
8. Never eat or drink in the laboratory.
9. Wash your hands before and after each activity in the lab.
10. Keep the work area free of any unnecessary items.
11. Wash all utensils thoroughly before and after each use.
12. Never smell or taste any chemicals unless instructed to do so by your teacher and the experiment instructions.
13. Do not experiment or mix chemicals on your own. Many chemicals in the lab are explosive or dangerous.
14. When using scissors or a scalpel, cut away from yourself and others.
15. When heating substances in a test tube, always point the mouth of the test tube away from yourself and others.
16. Clearly label all containers with the names of the materials you are using during the activity.
17. Report all accidents to the teacher immediately, including breakage of materials, chemical spills, and physical injury.
18. Do not pick up broken glass with your hands. Sweep up broken glass with a broom, and dispose of the glass in a container labeled for glass disposal.
19. Never return unused chemicals to their original containers. Follow your teacher's instructions for the proper disposal and cleanup of all materials prior to the end of the lab period.
20. Make sure all your materials are washed and put away, and your work area is clean, before leaving the lab.
21. Be certain that all Bunsen Burners, gas outlets, and water faucets are turned off before leaving the lab.

SAFETY SYMBOLS

Symbol	Description	Symbol	Description
	Eye Safety		Fire/Explosion Safety
	Clothing Protection		Electrical Safety
	Glassware Safety		Poison
	Sharp Objects		Animal Safety
	Heating Safety		Plant Safety

Eye Safety
- Wear your laboratory safety goggles when you are working with chemicals, open flame, or any substances that may be harmful to your eyes.
- Know how to use the emergency eyewash system. If chemicals get into your eyes, flush them out with plenty of water. Inform your teacher.

Clothing Protection
- Wear your laboratory apron. It will help to protect your clothing from stains or damage.

Glassware Safety
- Check glassware for chips or cracks. Broken, cracked, or chipped glassware should be disposed of properly.
- Do not force glass tubing into rubber stoppers. Follow your teacher's instructions.
- Clean all glassware and air-dry them rather than drying with a towel.

Sharp Objects
- Be careful when using knives, scalpels, or scissors.
- Always cut in the direction away from your body and from others who are nearby.
- Inform your teacher immediately if you or your partner is cut.

Heating Safety
- Turn off heat sources when they are not in use.
- Point test tubes away from yourself and others when heating substances in them.
- Use the proper procedures when lighting a Bunsen burner.
- To avoid burns, do not handle heated glassware or materials directly. Use tongs, test-tube holders, or heat-resistant gloves or mitts.
- For heating, use glassware that is meant to be used for that purpose.
- When heating flasks or beakers over the laboratory burner, use a ring-stand setup with a square of wire gauze.
- Use a water bath to heat solids.
- When heating with a laboratory burner, gently move the test tube over the hottest part of the flame.
- Do not pour hot liquids into plastic containers.

Fire/Explosion Safety
- Tie back long hair and roll up long sleeves when working near an open flame. Confine loose clothing.
- Do not reach across an open flame.
- Know the location and proper use of fire blankets and fire extinguishers.

Electrical Safety
- Be careful when using electrical equipment.
- Check all electrical equipment for worn cords or loose plugs before using.
- Keep your work area dry.
- Do not overload electric circuits.
- Be sure that any electrical cords are not in a place where someone can trip over them.

Poison
- Tie back long hair and roll up long sleeves when working with chemicals.
- Do not mix any chemicals unless directed to do so in a procedure or by your teacher.
- Inform your teacher immediately if you spill chemicals or get any chemicals on your skin or in your eyes.
- Never taste any chemicals or substances unless directed to do so by your teacher.
- Keep your hands away from your face when working with chemicals.
- Wash your hands with soap and water after handling chemicals.

Animal Safety
- Handle live animals with care. If you are bitten or scratched by an animal, inform your teacher.
- Do not bring wild animals into the classroom.
- Do not cause pain, discomfort, or injury to an animal.
- Be sure any animals kept for observations are given the proper food, water, and living space.
- Wear gloves when handling live animals. Always wash your hands with soap and water after handling live animals.

Plant Safety
- Use caution when collecting or handling plants.
- Do not eat or taste any unfamiliar plant or plant parts.
- Wash your hands with soap and water after handling plants.
- If you are allergic to pollen, do not work with plants or plant parts without using a gauze pollen mask.

DOING SCIENCE

Science is not simply a set of facts. It is really a kind of process—a way of learning about and understanding nature. Scientists know that their ideas about the universe are not fixed and certain. This uncertainty is not a sign that something is wrong with their work or their thinking. It comes partly from the fact that many of the things scientists study cannot be observed directly.

However, scientific uncertainty goes beyond that fact. It is built into science itself and is actually one of the great things about it. Science does not deal in absolute facts and absolute laws. No one is forced to accept scientific ideas just because a scientist has put them forth. Such ideas must be tested, challenged, and questioned. If these ideas turn out to be successful in describing the way the world works, they are accepted and used. Even then, however, scientists do not consider these ideas to be certainties. The process of science will continue, and the ideas will eventually develop or be replaced by other ideas that work even better.

There are many different methods and techniques that can be used in science. However, all truly scientific methods have certain things in common. They all involve an open-minded approach to nature and systematic ways of gathering and processing information, and of coming to conclusions. There is no single, correct step-by-step order in which scientific investigation must be carried out. However, the description below will give you a good idea of how the approach works in general.

OBSERVING

A scientist usually begins by noticing something about the world—perhaps the fact that a certain kind of plant (say, a bean plant) does not seem to grow equally well under all conditions. The initial observation is often followed up with more careful observations, perhaps involving measuring instruments. The scientist will wonder why the thing that is observed behaves the way it does, or perhaps wonder what factors affect its behavior. In the case of the plants, for example, he or she might wonder what factors cause bean plants to grow rapidly and remain healthy.

HYPOTHESIZING

Such observing and wondering often leads to an educated guess about what has been observed. Such an attempt to explain observations in terms of their causes is called a *hypothesis*. In the case of the bean plants, the scientist might make the hypothesis that rapid and healthy growth depends upon the plants' receiving light. A statement of the hypothesis might be that bean plants that are exposed to light grow better than those that are not exposed to light.

DESIGNING AN EXPERIMENT

A non-scientist might be content to stop with an educated guess and simply assume that it is correct. A good scientist assumes nothing, however. Instead, he or she plans a way to test hypotheses to see whether they will hold up and apply to actual new situations. A way of testing hypotheses and gathering more information is called an *experiment*. Although experiments are sometimes used simply to gather background information before any hypothesis is made, they generally follow the hypothesis step and serve as a check on the soundness of the hypothesis.

Experiments must be designed carefully if they are to be of use in answering questions. A good experiment permits observation of one particular factor that is a suspected cause of something observed. This factor whose effects are to be studied is called the variable. For example, the amount of light to which bean plants are exposed can serve as the variable in the experiment.

It is not enough simply to expose a plant to light and to observe what happens. The variable factor, light, must, as its name implies, be *varied*. For example, two bean plants might be observed. Plant A will be exposed to light. Because it contains the variable, the setup involving plant A will serve as the experimental setup. A different bean plant, B, will not be exposed to light. The B setup is called the control setup. It lacks the variable, light.

It would not make sense to treat the two plants differently in any respect other than the amount of light they are given. For example, if one plant were watered and the other were not, the effects observed could not be interpreted clearly. You would not know which effect, the amount of water or light, was responsible , or whether both were. Therefore, all factors other than the variable one are controlled, or kept the same, for the two setups.

COLLECTING AND ORGANIZING DATA

The experiment that has been designed is then carried out. Observations and measurements are carefully made and recorded. For example, data on the heights of the two plants over the following several weeks might be graphed, to permit easy comparison.

INTERPRETING DATA AND COMING TO A CONCLUSION

The organized data are then studied. For example, it might be found that plant A grew 8 centimeters per week, whereas plant B grew only 1 centimeter. Then a conclusion is drawn on this basis as to the correctness of the original hypothesis. In the example of our bean experiment, the conclusion might be that light does have a positive effect on the growth rate of bean plants.

COMMUNICATING RESULTS

Scientific findings are not of much use unless others are informed of them. Scientists thus communicate their findings to others. This can be done informally, but is usually done by means of written reports that are published in scientific journals. Such reports provide complete information on the experiments that were done. Thus, other scientists need not take the results on faith. They can analyze the design and conclusions of the experiment, or repeat the experiment themselves. This repeatability is a good check on the correctness of scientific conclusions.

In carrying out the procedures described in this manual, think carefully about what you are doing and why you are doing it. Keep in mind the hypothesis you have formulated and how the data you will gather will help to confirm or reject it. Work carefully and systematically and collect data accurately on the things you observe and measure. Take care analyzing the data and think about what the findings reveal. Finally, make conclusions based upon your findings. If you do all these things, you can congratulate yourself on acting as a real scientist does and you will learn a great deal in the process.

LABORATORY INVESTIGATION

1 THE ATMOSPHERE AND LIVING THINGS

Problem: *How have the atmosphere and living things interacted throughout Earth's history?*

INTRODUCTION

Background It has been proposed that Earth functions as if it were one huge organism. The parts of Earth consist of living things and environmental factors, existing in a balanced state. If one part changes, the other parts will also change to maintain this balance, as long as the changes are relatively small. However, large changes in one part of Earth can greatly alter the biosphere. The perception of Earth as one giant living being is highly controversial, but the idea that Earth maintains a balance between living and nonliving parts is widely accepted. This interaction between living and nonliving things can be observed in Earth's changing atmosphere. Life on this planet has adapted to a changing atmosphere. These life forms have, in turn, altered the atmosphere.

Goals In this investigation, you will plot the percentages of various gases present in Earth's atmosphere. You will then **construct a graph** depicting changes in the atmosphere through time, and **identify** major geological events on the same graph. You will then **interpret** the graph, relating the events to the atmospheric changes.

LAB WARMUP

Concepts A useful way to show change through time is to plot data on a graph. Scientists use graphs to visualize trends they cannot readily observe from raw data acquired in their research. A *cumulative graph* shows data of several different types. A cumulative graph is especially useful when showing the composition of something, or the amounts of different parts that make up the whole.

Suppose a scientist is studying how the bird population on a small island has changed over the last 150 years. To visualize the data, the scientist might use a cumulative graph (Figure 1.1). Notice that each portion of the bird population is built upon the previously graphed data. The blackbird percentages are graphed first, by plotting the actual percentages in the population. The bluebird percentages are built upon the blackbird. For the year 1900, for example, the percentage of bluebirds in the population was 50% minus 30%, or 20%. For the same year the percentage of redbirds in the population was 85% minus 50%, or 35%, and the remainder of the population (100% minus 85%, or 15%) was composed of other species of birds. For any given year, the sum of the percentages of all categories is always 100%.

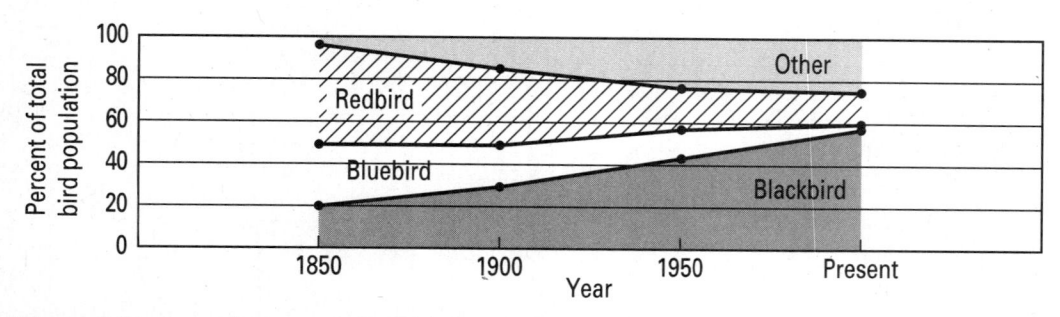

Figure 1.1 Cumulative graph of bird population on island

Review Section 1.3, The Atmosphere, should be completed before beginning this investigation. You should also understand the following terms before you perform this investigation.

atmosphere   organism   gas

Make a **prediction** about the outcome of this experiment and write it in the Lab Notebook.

MATERIALS (PER STUDENT)

- pencil
- colored pencils, markers, or crayons

PROCEDURE

1. Figure 1.2 illustrates the composition of Earth's atmosphere at different times in the planet's history. Plot the percentages of the gases on the graph provided in your Lab Notebook. Your graph should be a cumulative graph; use the graph from the Concepts section as a guide. First observe that the amount of carbon dioxide in the atmosphere 4.5 billion years ago was 80 percent. Plot that number in the appropriate space on the graph.
2. The next gas in the chart is nitrogen. Using a different color, plot a point 10 percentage points higher on the graph, at 80 + 10, or 90% of the total atmosphere. Proceed with the other gases in this manner. Use a different color to plot the points for each graph.
3. After all the points are plotted, connect the points plotted for each gas, producing a curve for each. This curve represents gradual change over time.
4. Color the areas representing different gases. For example, the area beneath the line drawn for carbon dioxide represents the proportion of carbon dioxide in the atmosphere. Shade this space with the color you used when plotting the points for carbon dioxide. Label all gases.
5. Figure 1.3 lists some important events in Earth's history. Write the events underneath your graph next to the times they occurred. The first event has been done for you.

Gas	Billions of years before present									
	4.5	4.0	3.5	3.0	2.5	2.0	1.5	1.0	0.5	Present
Carbon dioxide	80%	20%	10%	8%	5%	3%	1%	0.07%	0.04%	0.025%*
Nitrogen	10	35	55	65	72	75	76	77	78	78
Hydrogen	5	3	1	0.5	0	0	0	0	0	0
Oxygen	0	0	0	0	0	1	5	10	15	21
Other gases	5	42	34	26	23	21	18	13	7	1

* The carbon dioxide level of the atmosphere has risen to a current level of 0.04% due to human activities.

Figure 1.2 Composition of Earth's atmosphere from Earth's formation until present

Geological Event	Billions of Years Ago
Origin of Earth	4.5
Formation of oldest known bedrock	3.9
First evidence of organic matter in rocks	3.7
Photosynthesis evolves in plants	3.0
Limestone deposits become common	1.8
Many fossils of marine invertebrates	0.55
Earliest land plants	0.44
Earliest land animals	0.40
Dinosaurs dominant	0.17

Figure 1.3 Major geological events in Earth's history

Name: _____ Class: _____ Date: _____

LAB NOTEBOOK: INVESTIGATION 1

PREDICTION A correct prediction is that a cumulative graph will show that Earth's atmosphere has changed over time.

OBSERVATIONS

Evolution of Earth's atmosphere

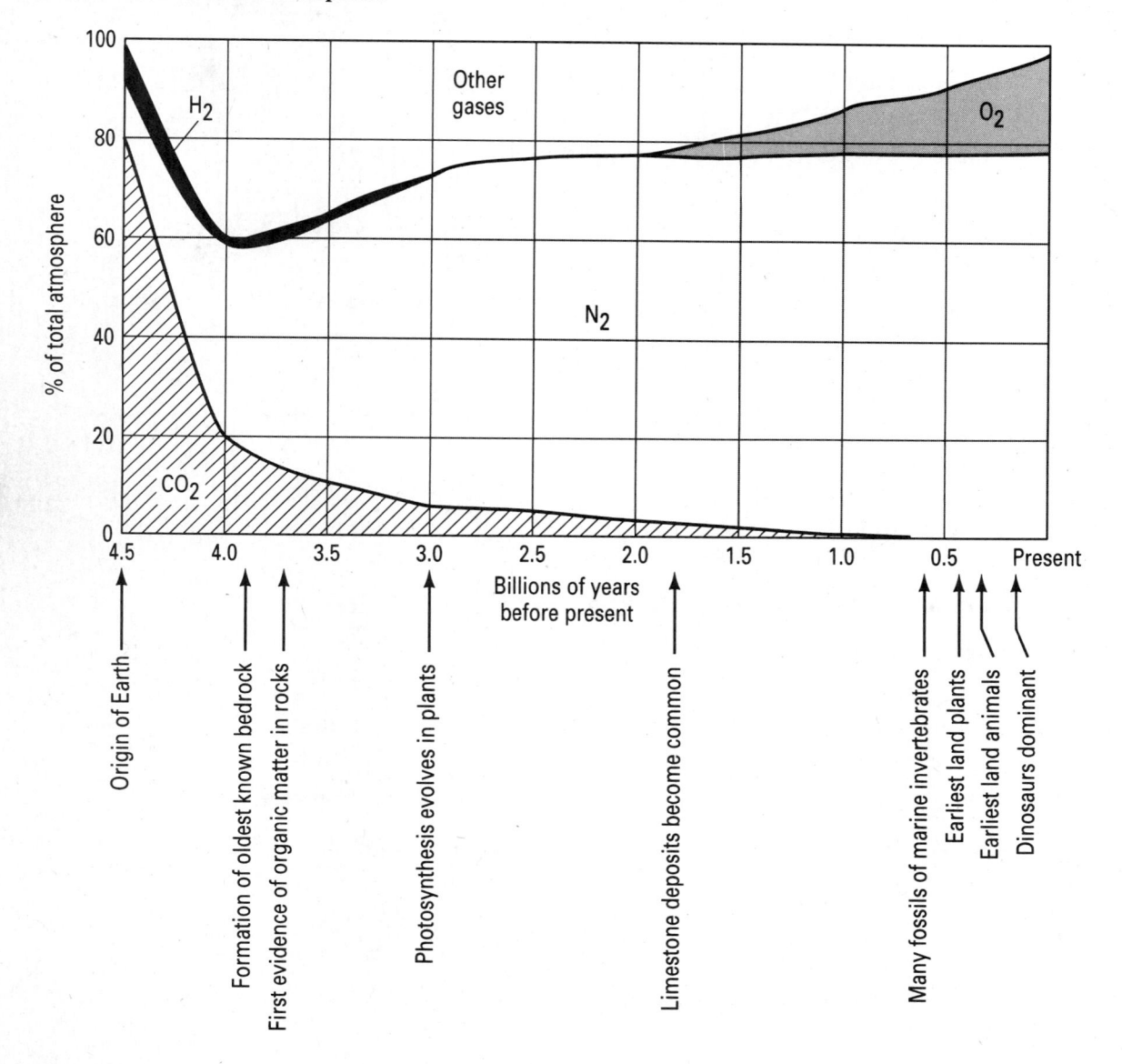

[Graph showing % of total atmosphere (y-axis, 0 to 100) vs. Billions of years before present (x-axis, 4.5 to Present). The graph shows CO_2 starting high at ~80% and decreasing rapidly, H_2 band decreasing from top, N_2 making up the bulk of the atmosphere through most of the timeline, Other gases, and O_2 increasing in recent times.]

Arrows along the x-axis indicate:
- Origin of Earth (4.5)
- Formation of oldest known bedrock
- First evidence of organic matter in rocks
- Photosynthesis evolves in plants (~3.0)
- Limestone deposits become common (~2.0)
- Many fossils of marine invertebrates
- Earliest land plants
- Earliest land animals
- Dinosaurs dominant

© Addison-Wesley Publishing Company, Inc. All Rights Reserved.

DATA ANALYSIS

1. How old is Earth?
   Earth is approximately 4.5 billion years old.

2. What gas has made up the largest portion of Earth's atmosphere for most of Earth's history?
   nitrogen

3. Which gas appeared in the atmosphere about the time when limestone deposits became common?
   oxygen

4. How does the appearance of photosynthetic plants relate to the increase in atmospheric oxygen and the decrease in carbon dioxide?
   Plants use carbon dioxide and give off oxygen during photosynthesis.

5. If the trends seen in the graph continue, how will Earth's atmosphere change in the next 500 million years?
   Levels of oxygen will continue to rise; levels of carbon dioxide and other gases (except nitrogen) will continue to drop.

CONCLUSION

1. **Predict** Limestone is composed of calcium carbonate ($CaCO_3$). Protists use free calcium ions (Ca^+) from the sea to form calcium carbonate shells. Using this information, explain why limestone deposits became common 1.8 billion years ago.
   Once free oxygen became available, protists evolved. The protists could bind calcium ions, forming shells which eventually settled to form limestone deposits.

2. **Compare** The atmospheres of Mars and Venus are composed primarily of carbon dioxide. Why is Earth's atmosphere different?
   Photosynthetic plants use carbon dioxide during photosynthesis and release oxygen into the atmosphere. Mars and Venus have no organisms to alter their atmospheres.

3. **Interpret** How do the changes in atmospheric composition throughout Earth's history illustrate the relationship between living and nonliving parts of the environment?
   The atmosphere (environment) and organisms inhabiting Earth interact, causing changes on Earth and in organisms.

EXTENSION

Research The increase in Earth's atmospheric levels of oxygen gas was due to life processes. The decrease in the amount of carbon dioxide in the atmosphere was also due to life processes. What was the cause of the elimination of hydrogen gas in Earth's atmosphere?

LABORATORY INVESTIGATION

2 SAMPLING A BIOTIC COMMUNITY

Problem: *How can the population sizes of various species of organisms be estimated using the quadrat method?*

INTRODUCTION

Background The biotic factors of an environment are all organisms found within that environment. *Organisms* are divided into five kingdoms: Monera (bacteria, blue-green algae), Protista (single-celled organisms such as amoebas), Fungi (including mushrooms and molds), Plantae (plants), and Animalia (animals). A group of organisms of the same species found within the same environment is called a *population*. Populations of different species sharing the same area and interacting with each other make up a *community*. One way of studying the interactions of organisms in a community is by taking an inventory of all the species in the area and comparing the sizes of their populations.

Goals In this investigation, you will **observe** the abiotic factors of a study site and **classify** the species of plants and animals in that site. You will then **measure** randomly chosen quadrats within the site and **count** the individuals of each population located within each quadrat. With this data you will **estimate** the sizes of plant and animal populations within the community.

LAB WARMUP

Concepts Before scientists can design an experiment, they must first make observations on which to base their hypotheses. Scientists have many different methods of collecting this data. The task of taking an inventory of the different kinds of organisms and their population sizes in an environmental site can be very difficult, especially if the area is teeming with life. Since it would be impractical, if not impossible, to count each individual organism in a large area, ecologists randomly choose small portions of the whole area and classify and count the organisms in each small portion. They can then estimate the size of each population in the larger community. This process is called the *quadrat method*.

The goal of the quadrat method is to estimate the population density of each species in a given community. Population density is the number of individuals of each species per unit area. Small square areas, called quadrats, are randomly selected to avoid choosing unrepresentative samples. Once the population densities for all quadrats are determined, the population size within the larger area can be estimated.

For example, if a 10 m × 10 m (100 m^2) site is being surveyed, three quadrats of a smaller size, perhaps 1 m × 1 m (1 m^2), might be selected at random. If the population densities of a particular species at the three quadrats are 10, 12, and 14 individuals per m^2, an average is taken [(10 + 12 + 14)/3 = 12 individuals per m^2]. That number is multiplied by the ratio of the larger area to the area of each quadrat (100 m^2/1 m^2) to calculate the estimated population size within the site (12 × 100 = 1200 individuals). This process is repeated for all species in the community.

Review Sections 2.2, Skills and Methods and 2.3, Environmental Science should be completed before beginning this investigation. You should also understand the following terms before you perform this investigation.

**population   community   abiotic factors   biotic factors   random sampling
quadrat   population density**

Make a **prediction** about the outcome of this experiment and write it in the Lab Notebook.

MATERIALS (PER GROUP)

- protective work gloves
- metric ruler or tape measure
- 16 stakes
- rubber mallet
- right triangle tool
- large ball of string
- scissors
- field guide books to local plant and animal species
- notepad for recording data

PROCEDURE

CAUTION: Wear protective work gloves when handling plants and animals. Be able to identify poisonous plants and alert your teacher beforehand of any allergies to plants or insects you may have.

1. Using the tape measure, mark off a square 10 m on each side and drive a stake into the ground at each corner. **CAUTION: Be careful not to injure yourself or others when using the mallet and stakes.** Use the right triangle tool or a notebook to make the square as precise as possible.

2. Loop the string around each of the four stakes to mark the boundaries of the study site, then cut the string and tie the ends. Be sure the string is taut.

3. Observe the abiotic factors of your site, such as whether the area is located in full sun or shade, or whether the soil is moist or dry. Record your observations in the Lab Notebook.

4. Take an inventory of the different kinds of plants and animals found in your site. Use field guide books or other references to identify the species you observe. When surveying animals such as insects or worms, look under rocks, on branches, and in the soil, trying your best not to disturb them.

5. Select at random an area within the site to be your first quadrat. To do this, close your eyes and toss a small object (rock, coin, etc.) into the square. Measure off a square 1 m × 1 m, making the point where the object landed the center of the square and making the sides of the quadrat parallel to the sides of the larger square. Use the stakes and string to make the quadrat. Again use a right triangle or notebook to make the square exact.

6. Record in the Lab Notebook the number of organisms of each plant or animal species within the quadrat. To count grass or very small insects, select three smaller (10 cm × 10 cm) squares at random, count the number of individual plants or insects in each smaller square, average the numbers, and multiply by 100 to get an estimate for the full quadrat.

7. Repeat steps 5 and 6 twice more to obtain data for two other quadrats within the site.

8. For each species, add the number of organisms found in all three quadrats and divide by three to calculate the average population density per square meter. Record the average population densities in the Lab Notebook.

9. For each species, multiply the population density by 100 to estimate the total number of organisms in the larger site. Record the estimated population sizes in the Lab Notebook.

Name: _____ Class: _____ Date: _____

LAB NOTEBOOK: INVESTIGATION 2

PREDICTION The quadrat method can be used to estimate population sizes within a biotic community.

OBSERVATIONS

Abiotic factors

Answers might include amount of sunlight, exposure to wind, moistness of the soil, etc.

Inventory of plant species   Inventory of animal species

Species	No. in Quad. 1	No. in Quad. 2	No. in Quad. 3	Average pop. density	Est. population size in site
Answers will vary.					

© Addison-Wesley Publishing Company, Inc. All Rights Reserved.

15

DATA ANALYSIS

1. Why was it necessary to close your eyes before choosing the quadrat?
   to ensure that the quadrat was chosen randomly

2. Which was the dominant plant species within the site? The dominant animal species?
   Answers will vary. The dominant species will be the most numerous.

3. How do your averages compare to the population densities of the individual quadrats? Were the populations spread out evenly over the site?
   Answers will vary. If a population is spread out evenly over the site, the average will be similar to the population density within each quadrat.

4. Did you observe any unusual features in one quadrat that were not found elsewhere on the site, such as an ant hill?
   Accept all logical responses.

CONCLUSION

1. **Infer** Compare your results with those of other groups. Did any abiotic factors contribute to similarities or differences in your population sizes? Explain.
   Answers will vary, but should show an understanding of how abiotic factors influence the composition of the community. For example, the moistness of the soil might affect plant population sizes within a site.

2. **Suggest** Using the quadrat method, how could you better estimate the actual population sizes of species within the site?
   Increasing the number of quadrats would produce a more accurate estimate.

3. **Compare** How is the sampling of animal populations more difficult than that of plant populations?
   Animals are mobile, and their movement in and out of the quadrat may affect the results. Also, they may be concentrated in very small areas, such as ant hills, which may not be sampled, or they may be located deep in the soil, where they cannot be counted.

EXTENSION

Predict How will the removal of the dominant plant species from the site affect the population sizes of the other organisms? Using Chapter 2 of your text as a guide, design a scientific experiment that could be used to test your hypothesis. Remember to include a control group and an experimental group, and change only one variable at a time.

LABORATORY INVESTIGATION

3 ABIOTIC FACTORS

Problem: *How do the abiotic factors of soil moisture, light intensity, and temperature affect communities of organisms?*

INTRODUCTION

Background The organisms of a community are constantly interacting. Changes in one population affect other populations in the community. Abiotic factors such as soil type, moisture, slope of the land, wind, light, and temperature also affect the composition of the community. Scientists who study communities of organisms benefit from observing changes in the abiotic factors of an ecosystem. Like biotic factors, abiotic factors change from season to season and from year to year. Organisms in the ecosystem either adapt to changing conditions or they die, thereby changing the composition of the community. In autumn, for example, as temperatures decrease, leaves fall from trees and some animals prepare for winter hibernation. Days become shorter and sunlight becomes less intense. Although many plants die during a winter freeze, they may have dropped seeds that sprout the following spring.

Goals In this investigation, you will **measure** three abiotic factors—moisture, light intensity, and temperature—in different areas. You will then **observe** the variety of plant species present in those areas and **infer** relationships between the biotic and abiotic factors.

LAB WARMUP

Concepts Scientists use many different tools to make measurements of abiotic factors. Litmus paper is used to determine the pH of soil and water. Anemometers measure wind speed in revolutions per second. Light meters measure light intensity in lumens or other units. Thermometers measure temperature in either degrees Fahrenheit or Celsius. All of these tools are used to make comparisons of the abiotic factors of an environment. In this lab you will use a light meter and thermometer to measure light and temperature.

When scientists design experiments, they change only one variable at a time, such as temperature. All other possible variables are kept constant. There must also be a control setup in the experiment, in which nothing has been changed.

Review Section 3.3, The Ecosystem, should be completed before beginning this investigation. You should also understand the following terms before you perform this investigation.

**population   community   abiotic factors   biotic factors   anemometer
thermometer   light meter   variable   control setup**

Make a **prediction** about the outcome of this experiment and write it in the Lab Notebook.

MATERIALS (PER CLASS)

- light meter
- thermometer
- 6 flags of different colors
- colored pencils

PROCEDURE

1. You will be assigned to a group that will measure one of the abiotic factors listed below. Follow the procedure for your assigned group. Write your data on the chalkboard so it may be copied by your classmates into their Lab Notebooks.

   Soil Moisture Choose 12 areas within the study site to measure the amount of moisture in the soil. Select six areas where the soil is very moist, and six where the soil is very dry. Make subjective measurements of soil moisture on a scale from 1 to 10. 1 indicates very dry soil; 10 indicates mud. Record your measurements in your Lab Notebook. Mark the point of highest soil moisture with a blue flag. Mark the point of lowest soil moisture with a green flag.

   Light Intensity Choose 12 areas within the study site to take light intensity measurements using the light meter. Select six areas where you think you will find the least intense light, and six where you think you will find the most intense light. Record your measurements in your Lab Notebook. Place an orange flag at the point of the most intense light and a yellow flag at the point of the least intense light.

   Temperature Choose 12 areas within the study site of take temperature measurements using the thermometer. Select six areas where you think you will find the lowest temperature, and six where you think you will find the highest temperature. Record your measurements in your Lab Notebook. Place a red flag at the point of highest temperature and a purple flag at the point of lowest temperature.

2. All groups should now survey the plant life around the areas marked by a flag. Observe leaf size and shape, plant height, and amount of leaves on each plant. Also observe whether plants are located close to each other or are sparsely distributed. Write in your Lab Notebook your general observations, such as "no vegetation," "grasses, weeds, and small shrubs," "trees only," etc., and any other observations you consider important.

3. Record the locations of your flags on the class site map provided by your teacher. Draw a circle at the estimated point of the actual flag. Write the name of the flag's color inside the circle, or use a colored pencil to fill it in.

4. Draw the types of vegetation surrounding the flags on the class map.

Name: _____ Class: _____ Date: _____

LAB NOTEBOOK: INVESTIGATION 3

PREDICTION <u>Abiotic factors of the environment affect the growth and distribution of different organisms.</u>

OBSERVATIONS

READINGS OF DIFFERENT AREAS OF STUDY SITE

Area	Soil moisture	Light intensity	Temperature
1			
2	**Answers will vary.**		
3			
4			
5			
6			
7			
8			
9			
10			
11			
12			

SURVEY OF PLANT LIFE LIVING IN ABIOTIC EXTREMES

Flags	Abiotic extreme	Observations
Blue	highest soil moisture	**Answers will vary.**
Green	lowest soil moisture	
Orange	most intense light	
Yellow	least intense light	
Red	highest temperature	
Purple	lowest temperature	

© Addison-Wesley Publishing Company, Inc. All Rights Reserved.

DATA ANALYSIS

PART A

1. Were there any areas where two or more flags of different colors were found together? If so, which flags were found together?
   Answers will vary. For example, the flag marking the area of the highest temperature might be found next to the flag marking the area of most intense light.

2. What types of plants were found around the flag with the highest soil moisture? Lowest soil moisture?
   Answers will vary. Mosses might be found in the area of high soil moisture. The area of low soil moisture may be surrounded by grass.

3. What types of plants were found around the flag with the most intense light? Least intense light?
   Answers will vary.

4. What types of plants were found around the flag with the highest temperature? Lowest temperature?
   Answers will vary, but will probably be similar to those for Question 3.

CONCLUSION

1. **Predict** What kind of vegetation would you expect to find in an area of relatively low light, high soil moisture, and low temperature?
   Answers will vary. Assuming that these conditions are caused by shade, there would be trees, possibly broad-leafed plants, and very little grass.

2. **Generalize** How do the abiotic factors of soil moisture, light intensity, and temperature affect the quantity and types of vegetation in an area? Support your answer with your lab results.
   Answers will vary depending on students' results. Broad-leafed plants will generally be found in areas of relatively low temperature, low light intensity, and high soil moisture. Grasses will generally be found in the opposite situation.

EXTENSION

Hypothesize Design an experiment to test the effect of varying temperature on grass seedling growth. Choose three different species of grass and expose planted seeds of each species to three different temperatures for a few days. (Place seedlings in freezer, refrigerator, and at room temperature.) Do different temperatures affect all species in the same way?

LABORATORY INVESTIGATION

4 OWL PELLETS

Problem: *What information about an owl's diet and role in the environment can be learned from an owl pellet?*

INTRODUCTION

Background   The owl, like many other carnivorous birds, swallows its prey whole. Many parts of an animal are not digestible, such as hair, feathers, bones, teeth, and the hard outer shells of insects. The owl's digestive system allows it to store these indigestible parts while letting the digestible parts pass to the intestines. The owl then regurgitates the unwanted parts in the form of a pellet. An owl pellet is a roundish mass that is covered with fur and sometimes feathers from its prey. Fresh specimens are shiny and coated with mucus. Bones and other hard remains are located within the interior of the pellet. The pellet can provide evidence of the owl's dietary habits and role in its environment.

The owl plays a role in limiting the population size of its prey. Of the animal species common to an owl's diet, the prey species that is most abundant in the area will be the species most likely to be captured and consumed by the owl. This limits the population size of the herbivore, which in turn protects the supply of plants upon which the prey feeds. Farmers especially appreciate barn owls, which keep populations of crop-eating rodents under control.

Goals   In this investigation, you will **observe** the external features of an owl pellet. You will then **reconstruct** and **identify** the skeletons of the prey contained in the pellet, and count the total number of organisms found. Finally, you will **infer** the owl's role in its environment and its place in the food web.

LAB WARMUP

Concepts   A *dichotomous key* is one tool scientists use to identify organisms. Each step of the key has two descriptions, and the organism being identified fits one of the two descriptions. Next to each description, the key either gives the name of the organism or directs the person to another step. The key is followed until the animal is identified.

Review   Sections 4.1, Roles of Living Things, and 4.2, Ecosystem Structure, should be completed before beginning this investigation. You should also understand the following terms before you perform this investigation.

predator   producer   consumer   carnivore   herbivore   dichotomous key

Make a **prediction** about the outcome of this experiment and write it in the Lab Notebook.

MATERIALS (PER GROUP)

- barn owl pellet
- metric ruler
- glass jar with lid
- water
- dishwashing liquid

- 2 pieces of cheesecloth cut into 6-cm squares
- bowl
- 2 sheets of white paper
- forceps, probe, or toothpicks
- field guides to mammals and birds
- books of animal skeletons

PROCEDURE

Wash your hands thoroughly after completing this lab.

1. Examine the outside of the owl pellet. Measure and record its length and width in centimeters. Describe external features in the Lab Notebook.
2. To investigate the interior of the pellet you must soften it by soaking it in water. Fill the jar halfway with water. Add a drop of dishwashing liquid and the pellet to the jar. Close the jar and shake gently for about 30 seconds. Let the jar stand for five minutes and shake again. When the pellet has fallen apart, pour the contents of the jar through two layers of cheesecloth. Collect the strained liquid in a sink or bowl.
3. Place the contents of the pellet on a piece of white paper. Pick bones, teeth, insect parts, and any other prey evidence out of the fur. Use probes, toothpicks, or forceps as necessary, but proceed carefully to avoid crushing any small bones. Put the animal remains on a second piece of paper. Discard the fur and other soft matter in a safe manner.
4. Label the skulls by number, and identify them. Use the key below to identify mammal skulls found in the owl pellet. You do not need the mandible (lower jaw) to identify skulls using the key. Use the pictures of skulls in Figure 4.2, as well as any field guides you may have, to aid in identification. Record the species of each skull in your Lab Notebook.

Does the animal have...	Then...
1. a) 3 or fewer teeth on each side of its upper jaw?	go to 2.
b) at least 9 teeth on each side of its upper jaw?	go to 3.
2. a) 2 biting teeth on its upper jaw?	go to 4.
b) 4 biting teeth on it upper jaw?	it's a rabbit.
3. a) a skull length of 23 mm or less and brown teeth?	it's a shrew.
b) a skull length of more than 23 mm and 44 teeth?	it's a mole.
4. a) the roof of its mouth extending past the last molar?	go to 5.
b) the roof of its mouth not extending past the last molar?	go to 6.
5. a) a skull length of 22 mm or less?	it's a house mouse.
b) a skull length of more than 22 mm?	it's a rat.
6. a) flat molars?	it's a meadow vole
b) rounded molars?	it's a deer mouse.

Figure 4.1 Key to mammals likely to be found in owl pellets

5. Try to match other bones to each skull. Using a book of animal skeletons, identify each bone and record its name under its matching skull in the Lab Notebook.
6. Use a field guide or other reference to find the likely habitat and diet of each species found in the pellet. Record the information in the Lab Notebook.

Shrew

Deer mouse

Mole

House mouse

Meadow vole

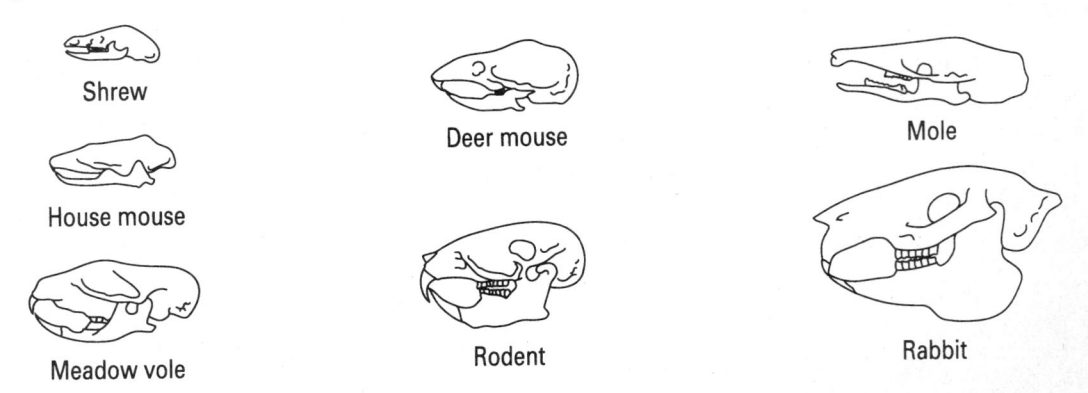

Rodent

Rabbit

Figure 4.2

Name: _____ Class: _____ Date: _____

LAB NOTEBOOK: INVESTIGATION 4

PREDICTION **A correct prediction is that an owl pellet will give valuable information about the dietary habits of the owl.**

OBSERVATIONS

OBSERVATIONS OF OWL PELLET

Length _____ (cm)   Width _____ (cm)

External features _____

_____

_____

_____

_____

ANIMAL REMAINS FOUND IN OWL PELLET

	Skull 1	Skull 2	Skull 3	Skull 4	Skull 5
Species					
Bones found					
Likely habitat					
Likely diet					

Use another sheet of paper if more than five skulls are found.

DATA ANALYSIS

1. How many individual animals did you find in the owl pellet?
   Answers will vary. Students will probably find between four and seven individual skeletons.

2. Were the animals of different species? If so, how many species were represented in the pellets?
   Answers will vary. Check student identifications of skulls.

3. Did you find any evidence of organisms other than small mammals in the owl pellet?
   Answers will vary. Students may find remains of birds, insects, and possibly (but not likely) amphibians, reptiles, and fish.

4. What types of bones are most prevalent in the pellet? What bones are least likely to be found in an owl pellet?
   Skulls, limb bones, ribs, and teeth are most likely to be found. Very small and fragile bones are least likely to be found.

5. What are the habitats of a barn owl's prey? Are they similar to the habitat of the owl?
   The barn owl is found in wooded areas, old farm buildings, and parks. Its prey is found in similar places.

CONCLUSION

1. **Predict** Other birds also form pellets. What would you expect to find in the pellet of a seagull?
   Since gulls live near the ocean, the pellet probably will contain remains of fishes, crustaceans, and insects.

2. **Classify** Describe the role of the owl in its environment and its place in the food web. Is it an herbivore or a carnivore? Is it a producer, a primary consumer, or a secondary consumer? Support your response with your lab results.
   The owl's role as predator makes it a carnivore (animal remains were found in the pellet). It is a secondary consumer because it eats animals (primary consumers) that eat plants (producers). If students find carnivorous animals in the pellet, they should classify the owl as a tertiary consumer. Students should describe the diet of the animals found in the pellet to support their classification.

EXTENSION

Model Reconstruct and mount the skeletons of animals found in the pellet for use in comparative anatomy. Place the bones in a beaker filled with peroxide water (use half commercial 10%-peroxide and half water). Leave the bones in this solution for 24 hours. Rinse the bones with fresh water and allow them to dry. Reassemble the bones, using white glue to affix them to a piece of poster board.

LABORATORY INVESTIGATION

5 THE OXYGEN CYCLE

Problem: *How is the oxygen cycle demonstrated in a closed ecosystem?*

INTRODUCTION

Background Earth's atmosphere had no free oxygen when it formed 4.5 billion years ago. Due to the evolution of photosynthetic plants and the release of oxygen during photosynthesis, Earth's atmosphere is now 21 percent oxygen. Oxygen is present in the atmosphere, is dissolved in water, and is bonded to other elements in molecules, such as water, carbon dioxide, and glucose.

Like water, nitrogen, and carbon, oxygen is constantly cycled through an ecosystem. During photosynthesis, plants use carbon dioxide, water, and sunlight to make glucose ($C_6H_{12}O_6$). Free oxygen is produced during this process and is released into the atmosphere by the plants. The oxygen, together with glucose, is used by animals in turn for cellular respiration, a process that produces energy for cellular functions. Plants also undergo cellular respiration. The equations below summarize the processes of photosynthesis and cellular respiration. Notice that the products of photosynthesis are used in cellular respiration, and vice versa.

Photosynthesis:

light energy + CO_2 + H_2O → $C_6H_{12}O_6$ + O_2

Cellular respiration:

$C_6H_{12}O_6$ + O_2 → CO_2 + H_2O + chemical energy

Goals In this investigation, you will **model** eight closed ecosystems, each consisting of a plant, an animal, and a water solution in a test tube. You will then **observe** the conditions of the plants and animals and any changes in the color of the solutions. You will **relate** these observations to the oxygen cycle and then **deduce** the oxygen requirements of organisms in a closed ecosystem.

LAB WARMUP

Concepts A chemical indicator is a substance that changes color when the pH or other characteristic of a material changes. The *pH* of a solution is related to the concentration of hydrogen ions (H^+) in that solution. The pH scale ranges from 0 to 14, with neutral substances having a pH of 7. A pH lower than 7 signifies that the solution contains a higher concentration of hydrogen ions and is acidic. A pH higher than 7 signifies that the solution contains a lower concentration of hydrogen ions and is basic. *Litmus (or pH) paper* is used to show the pH of a particular substance. The paper is placed into a solution, and the resulting color of the paper indicates the pH of the solution.

Chemical indicators can show changes in the amount of dissolved carbon dioxide and dissolved oxygen in a solution, since these levels are related to pH. A decrease in pH corresponds with an increased carbon dioxide level. An increase in pH corresponds with an increased oxygen level. *Bromothymol blue (BTB)* and *neutral red (NR)* are two indicators that will be used in this investigation. Bromothymol blue is a chemical indicator that turns green or yellow with decreasing pH, thus showing an increase in the amount of carbon dioxide in a solution. Neutral red changes to yellow with increasing pH, thus showing an increase in oxygen.

Review Section 4.4, Cycles of Matter, should be completed before beginning this investigation. You should also understand the following terms before you perform this investigation.

photosynthesis   cellular respiration   indicator   closed ecosystem   pH

Make a **prediction** about the outcome of this experiment and write it in the Lab Notebook.

MATERIALS (PER GROUP)

- 8 culture tubes with caps
- marker
- masking tape
- litmus (pH) paper
- pond or dechlorinated water
- 2 droppers
- bromothymol blue solution
- neutral red solution
- 8 sprigs of *Elodea*
- 4 snails
- test-tube rack
- light source

PROCEDURE

1. Label the test tubes *1A*, *1B*, *1C*, *1D*, *2A*, *2B*, *2C*, and *2D*. Also label the tubes with your name and the date. **CAUTION: Be careful not to cut yourself when handling glassware. Wear your laboratory apron and safety goggles.**

2. Using litmus paper, measure the pH of the pond water. Record this in the Lab Notebook.

3. Fill the tubes with pond water, leaving room for the addition of indicators, snails, and/or plants.

4. Add 5 drops of bromothymol blue solution to tubes 1A, 1B, 1C, and 1D. Add 5 drops of neutral red solution to tubes 2A, 2B, 2C, and 2D. Mix well by shaking the tubes gently after capping them. **CAUTION: Bromothymol blue and neutral red will stain skin and clothing and may irritate the skin. Avoid getting indicators on your skin or in your mouth.**

5. Add two healthy sprigs of *Elodea* to tubes 1A, 1C, 2A, and 2C.

6. Add one snail to tubes 1A, 1B, 2A, and 2B. **CAUTION: Be careful not to harm the snail when placing it in the tube.**

7. Measure the pH of each tube and record your data in the Lab Notebook.

8. Cap all the tubes tightly and place them in a test-tube rack near a bright light. Use sunlight if possible. Do not let the tubes get too hot.

9. Wait 24 hours and observe the color and pH of the solution and the status of the organisms. Record these observations in the Lab Notebook. Observe the tubes once a day for the next four days, recording your observations.

10. Return the snails to your classroom aquarium, or give them to your teacher so he/she may return them to the pond.

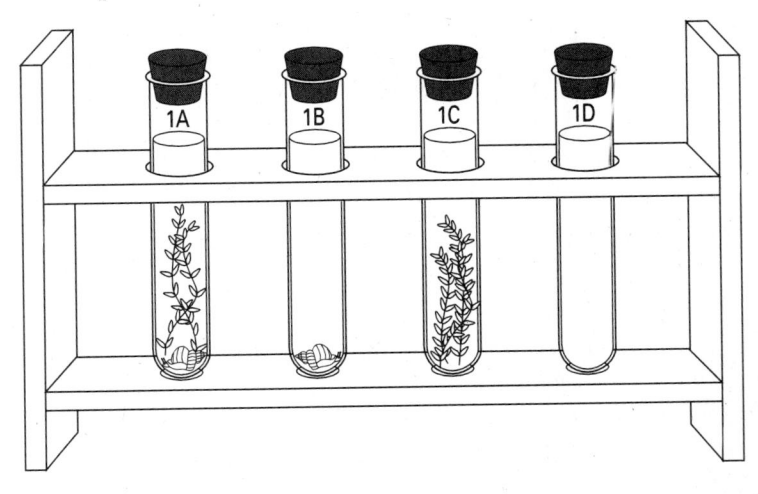

Figure 5.1 Set up for 1A-1D

Name: _____ Class: _____ Date: _____

LAB NOTEBOOK: INVESTIGATION 5

PREDICTION <u>A correct prediction is that an animal will survive in a closed ecosystem that includes a plant because it uses the oxygen produced by the plant during photosynthesis.</u>

OBSERVATIONS

pH of pond water = <u>Data will vary.</u>

CLOSED ECOSYSTEMS WITH BROMOTHYMOL BLUE

	Tubes			
	1A BTB + Elodea + snail	**1B** BTB + snail	**1C** BTB + Elodea	**1D** BTB only
Day 1				
Day 2	pH may decrease, color may change	pH will decrease; color will change	pH may decrease; color may change	pH and color will not change
Day 3				
Day 4				

CLOSED ECOSYSTEMS WITH NEUTRAL RED

	Tubes			
	2A NR + Elodea + snail	**2B** NR + snail	**2C** NR + Elodea	**2D** NR only
Day 1				
Day 2	pH may increase, color may change	pH and color will not change	pH will decrease; color will change to yellow or orange	pH and color will not change
Day 3				
Day 4				

DATA ANALYSIS

1. What is the purpose of tubes 1D and 2D?
   These are the control tubes of the experiment. They show the initial colors of the solutions without the addition of snails or *Elodea* so comparisons can be made.

2. Why is it important to cap the tubes tightly?
   Since this experiment represents a closed ecosystem, its oxygen cycle must progress without the addition of oxygen from outside the tubes.

3. Why is it necessary to place the tubes near a light source?
   Light energy is necessary for photosynthesis to occur.

4. Based on your observations, in which tube(s) did the level of carbon dioxide increase? In which tube(s) did the level of oxygen increase?
   Answers will vary. A BTB color change indicates an increase in carbon dioxide level. A neutral red color change indicates an increase in oxygen level.

CONCLUSION

1. **Predict** What would eventually happen to the snails in tubes 1B and 2B if the experiment was allowed to continue indefinitely? Explain.
   The snails would die from lack of oxygen. Since the *Elodea* is absent in these tubes, there is no continuous source of oxygen.

2. **Infer** What gas did the snail in tubes 1A and 2A produce that was used by the plant? For what process was it used?
   The snail produced carbon dioxide, which was used by the plant for photosynthesis.

3. **Integrate** In a closed ecosystem, could animals survive without plant life? Could plants survive without animal life? Explain.
   Animals could not survive without the release of oxygen by photosynthetic plants. Plants could survive without animals because they produce both oxygen (during photosynthesis) and carbon dioxide (during cellular respiration).

EXTENSION

Control Variables How important is the light source to the oxygen cycle? Can the cycle occur without this form of energy? Repeat the experiment, omitting the neutral red and adding 10 drops of bromothymol blue to all eight tubes. Place tubes 1A, 1B, 1C, and 1D near the light source, and place tubes 2A, 2B, 2C, and 2D in the dark. Record your observations each day. What difference does the light make in the oxygen cycle? Explain the importance of sunlight for most ecosystems.

LABORATORY INVESTIGATION

6 ESTIMATING POPULATION SIZE

Problem: *How can the population size of a mobile organism be measured?*

INTRODUCTION

Background The best way to measure the size of a population is to count all the individuals in that population. When determining the population sizes of trees or other relatively immobile organisms, this method is practical. If the organism is mobile, however, such as a fish, counting every individual would be difficult. Some individuals might be counted twice or not at all, since the experimenter would not know which fish had been counted and which had not. Ecologists use a method called the *Lincoln Index* to estimate the size of a population of mobile organisms.

Goals In this investigation, you will **model** a population of mobile organisms, capture and mark a sample of the population, and then capture a second sample. You will then **estimate** the size of the model population using the Lincoln Index. The accuracy of the Lincoln Index will be **inferred** by counting the model population.

LAB WARMUP

Concepts To use the Lincoln Index, scientists capture a sample of the population they want to measure. They mark these individuals and release them. After waiting several days, the scientists return and capture another sample. Some of the individuals in the second sample will carry the mark from the first sample.

The scientists then use the following formula to estimate the size of the population:

$$P = \frac{N_1 \times N_2}{R}$$

where **P** is the total size of the population
N_1 is the size of the first sample
N_2 is the size of the second sample
R is the number of marked individuals recaptured in the second sample

The Lincoln Index makes several assumptions that must be met if the estimate is to be accurate. These assumptions are:
- The population of organisms must be closed, with no immigration or emigration.
- The time between samples must be very small compared to the life span of the organism being sampled.
- The marked organisms must mix completely with the rest of the population during the time between the two samples.

Review Section 5.3, Populations, should be completed before beginning this investigation. You should also understand the following terms before you perform this investigation.

population   immigration   emigration   Lincoln Index   habitat

Make a **prediction** about the outcome of this experiment and write it in the Lab Notebook.

MATERIALS (PER GROUP)

- large, opaque jar with a tight lid
- several handfuls of dry beans
- red felt-tip pen
- blue felt-tip pen

PROCEDURE

1. The jar with the lid represents the habitat of your model population. Add several handfuls of beans to the habitat. The beans represent the organisms in the model habitat. *Note: Do not count the exact number of beans until the end of this investigation.*

2. Remove a small handful of beans from the model habitat. This handful will be your first sample. The sample should be at least 20 beans, but less than half of the total population.

3. Use the red felt-tip pen to mark each of the beans in the first sample. Make the marks big enough to be easily seen. Count the beans as you mark them. When you have finished marking and counting, let the red marks dry for a minute. Record the size of the first sample under N_1 in the Lab Notebook.

4. After the ink on the beans has dried, place them back into the model habitat and cover the jar. Each member of the group should shake the jar well to thoroughly mix the beans.

5. Remove the lid from the model habitat. Without looking, one member of the group should remove another handful of beans. The sample should be about the same size as the first sample. Count the total number of beans in this sample, then count the number of beans in the sample that are marked with red. Record these numbers in the Lab Notebook under N_2 and R Trial 1, respectively.

6. Return the beans to the jar, cover and shake well.

7. Two members of the group who have not taken a second sample should repeat steps 5 and 6. Be sure to mix the beans well between samples. Record the data in the Lab Notebook under N_2 and R, Trials 2 and 3.

8. Repeat this procedure, this time ignoring the red pen marks and using blue pen instead. Record your data in the second data table in the Lab Notebook.

9. Count the total number of beans in the jar. Work together with the others in your group. Record this number in both data tables in the Lab Notebook.

Name: _____ Class: _____ Date: _____

LAB NOTEBOOK: INVESTIGATION 6

PREDICTION _A correct prediction is that population size estimated by the Lincoln Index_
method is similar to the actual population size.

OBSERVATIONS

EXPERIMENT 1: RED PEN

N_1 (number of beans, first sample)	
N_2 trial 1 (number of beans, second sample)	**Data will vary.**
R trial 1 (number of beans marked in second sample)	
N_2 trial 2 (number of beans, second sample)	
R trial 2 (number of beans marked in second sample)	
N_2 trial 3 (number of beans, second sample)	
R trial 3 (number of beans marked in second sample)	
Average N_2	
Average R	
Actual number of beans in habitat	

EXPERIMENT 2: BLUE PEN

N_1 (number of beans, first sample)	
N_2 trial 1 (number of beans, second sample)	**Data will vary.**
R trial 1 (number of beans marked in second sample)	
N_2 trial 2 (number of beans, second sample)	
R trial 2 (number of beans marked in second sample)	
N_2 trial 3 (number of beans, second sample)	
R trial 3 (number of beans marked in second sample)	
Average N_2	
Average R	
Actual number of beans in habitat	

© Addison-Wesley Publishing Company, Inc. All Rights Reserved.

DATA ANALYSIS

1. Average the values for N_2 and R across the three trials in each experiment. Record the average values in the appropriate places in the Lab Notebook.
   Check student calculations for accuracy.

2. Use your data to estimate the size of the population of mobile organisms in the model habitat. Use the Lincoln Index in your calculations:

   $$P = \frac{N_1 \times N_2}{R}$$ Use the averages you calculated for N_2 and R.

   Averages will vary.

3. Compare population estimates calculated with the Lincoln Index to the actual size of the population. How do the estimates and the actual number compare?
   The estimated and actual population sizes should be similar.

CONCLUSION

1. **Interpret** Why did your estimates differ from the actual number of individuals in the model habitat? Discuss some factors that might affect the accuracy of your estimates.
   The estimates differed because they were based on random samples of the total population. The estimates may be affected by how well the jar was shaken, whether the beans were chosen randomly, and whether approximately the same number of beans was removed and counted each time. Larger samples will yield better averages.

2. **Infer** Compare your results with the results of other groups. Which group's estimates were most accurate? Compare the sample sizes of the groups. Is there an inference you can draw about the size of the samples and the accuracy of the Lincoln Index?
   The most accurate estimates will closely approximate the actual number of beans in the jar. The Lincoln Index is more accurate when large sample sizes are counted.

3. **Deduce** Why is it important that the time between the first and second samples represents a short time within the organism's life span?
   This method helps ensure that the marked organisms from the first sample will still be alive when the second sample is counted.

EXTENSION

Predict How will immigration and emigration affect population size estimates calculated with the Lincoln Index? Test your prediction by repeating the experiment. This time, however, remove 75 random beans from the jar before you take the second sample and replace them with 75 new, unmarked beans. The removal of beans models the emigration of individuals from the population. The addition of new beans models the immigration of individuals into the population. Complete the experiment. How did the immigration and emigration affect your estimate?

LABORATORY INVESTIGATION

7 MODELING EVOLUTION

Problem: *How can we model the mechanisms of evolution?*

INTRODUCTION

Background Ecosystems are constantly changing. Only a certain number of individuals from a given species can live in one habitat because resources such as food, water, space, and air are limited. A change in the environment may threaten the life of an individual unless it has some special trait that enables it to adapt to the environmental change. Such a trait is produced by mutation, a random change in the genetic material of an organism. The trait may give an individual a better chance to survive and reproduce than similar individuals that do not possess that trait. The genes for the trait will be passed to the next generation of organisms, and the trait may then become more frequent in the population over time. This process of natural selection, proposed by English naturalists Charles Darwin and Alfred Russell Wallace in 1858, is one process that influences evolution, the change in a population of organisms through time.

Darwin and Wallace were not the first people to propose an explanation for the changes in species through time. Jean Baptiste de Lamarck (1744–1829) argued that the earth was very old and had undergone gradual changes over time. Organisms had to change in their lifetime to better cope with their environment. Organisms that acquired these adaptations passed them on to their offspring, and gradually the species as a whole changed. Lamarck's ideas were inaccurate; changes acquired in one's lifetime cannot be passed on to the next generation. Lamarck and Darwin were missing a piece of the puzzle; it was not until the discovery of the principles of inheritance (genetics) and the source of new variants in a population (mutations) that evolution was truly understood.

Goals In this investigation, you will **model** the mechanisms of evolution, including random mutations that cause visible changes in an organism, and the process of natural selection.

LAB WARMUP

Concepts Mutations are random. They cause changes in proteins that control cellular functions and make up cell structures. Mutations are usually harmful or neutral; they very rarely give an individual an advantage over other individuals in a population. In natural selection, certain individuals of a population that possess a unique trait are more successful than the rest of the population in passing on their genes. Natural selection requires variation in a population upon which to operate. One way this variation is provided is by mutations.

Review Section 5.2, Evolution and Adaptation, should be completed before beginning this investigation. You should also understand the following terms before you perform this investigation.

mutation   natural selection   evolution   variation

Make a **prediction** about the outcome of this experiment and write it in the Lab Notebook.

MATERIALS (PER GROUP)

- paper
- scissors
- paper bags
- graph paper
- pencil

PROCEDURE

1. Copy the directional instructions below on a sheet of paper, and cut them into separate strips. (Your teacher may have already done this for you.) Put the strips in a paper bag.

one left	two left	no change
one right	two right	no change
one up	two up	repeat last step
one down	two down	reverse last step
diagonal right-up	diagonal left-up	
diagonal right-down	diagonal left-down	

2. Fill in the center square of the graph paper. This square will be your starting point.
3. Shake the strips of paper in the bag and, without looking, select one strip of paper from the bag.
4. Follow the direction on the strip and darken the square you come to on the graph paper.
5. Repeat steps 3 and 4 a total of 20 times, using your last darkened square as your starting point. The resulting figure represents an organism. Figure 7.1 shows a sample grid.
6. Circle the last square darkened. Compare your organism with the other organisms in your group. Since all students create their organisms independently, the organisms within each group should be different.
7. In your organism's habitat, there is a certain pressure that allows only those organisms possessing a specific trait to survive. One person in the group, acting as natural selection, will define and announce what that trait will be (choose one from below or define your own). Write that trait in your Lab Notebook. Possible advantageous traits in the habitat include: large central gap, long diagonal (or straight) chain, many right angles, many small gaps, overall rectangular or circular appearance.
8. Only those organisms within your group that show the advantageous trait will survive. Mark an "X" through those organisms that do not survive. Have the person representing natural selection make photocopies of those organisms so that each member of the group will have a copy of one surviving organism. For example, if two organisms survive, make two copies of one and three of the other for a group of five. Distribute the copies to the members of your group. Label the "parent" organism with the number of its generation and set it aside.
9. Using the photocopy of the surviving organism and the circled square as the starting point, repeat steps 3 through 8. The squares you darken this time, in addition to the squares darkened in the previous generations, will represent another organism, a descendant of the organism you previously created.
10. Repeat steps 3 through 9 until each student of the group has had a chance to play the role of natural selection. Your final drawing will show the history of changes, or mutations, that occurred among related organisms over a period of six generations. Observe the differences between the original organism and its descendants and record this information in your Lab Notebook.

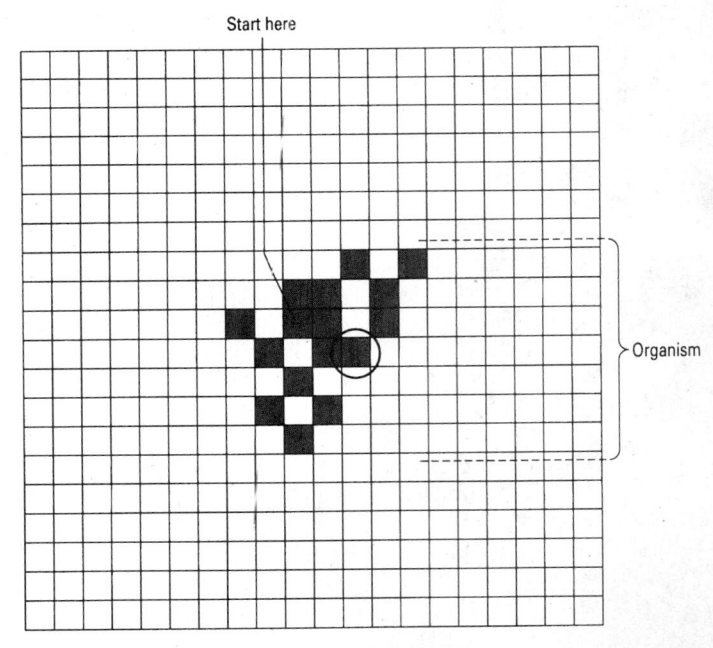

Figure 7.1 Sample grid

Name: _____ Class: _____ Date: _____

LAB NOTEBOOK: INVESTIGATION 7

PREDICTION **A correct prediction is that random mutations, natural selection, and evolution can be represented using a nonbiological model.**

OBSERVATIONS

TRAITS PROVING ADVANTAGEOUS TO THE ORGANISM IN ITS HABITAT

Generation
1. _____
2. _____ **Answers will vary.** _____
3. _____
4. _____
5. _____

OBSERVATIONS OF CHANGES IN ORGANISMS THROUGH SIX GENERATIONS

_____
_____
_____
_____
_____
_____
_____
_____
_____
_____
_____
_____
_____
_____
_____
_____

DATA ANALYSIS

1. How are mutations represented in this lab?
   Mutations are represented by the random changes delegated by the directional slips of paper, resulting in certain squares being darkened.

2. Why were you instructed not to look when selecting a directional slip of paper?
   The "mutations" needed to be randomly occurring to represent actual mutations.

3. How are organisms represented in this lab?
   Each student's drawing after completing 20 "mutations" is one "organism."

4. Are the "organisms" of each generation similar to or different from other "organisms" of the same generation within the group? That is, how much variation is present within each generation?
   The organisms should be different from each other, since each student randomly chooses mutations.

CONCLUSION

1. **Apply** Did you purposefully give the organisms you drew certain adaptive traits for survival in their particular habitat? Explain.
   Students should point out that the mutations were random and that the adaptive trait was not mentioned until after the drawings were made, so that it was only by chance that some of the organisms had the trait and survived.

2. **Model** How was the selection of advantageous traits by members of the group similar to the process of natural selection?
   The selection of advantageous traits determined which organisms survived and reproduced, just as natural selection does.

3. **Generalize** How are mutations important in the process of evolution?
   Mutations provide new traits that may be advantageous. These traits may be selected for within a population, causing changes in the population over time.

EXTENSION

Research A particular character trait originating from a mutation does not necessarily have to be advantageous to increase in frequency in a population. Research genetic drift and explain its relationship to natural selection.

LABORATORY INVESTIGATION

8 PREDATOR/PREY INTERACTION

Problem: *How are changes in the population sizes of predator and prey organisms related?*

INTRODUCTION

Background Organisms interact in many different ways. Some of the interactions have to do with feeding patterns. These feeding relationships make up what are called food chains.

Predator organisms feed upon other organisms, called prey. The predators depend on the populations of these prey organisms. The number of predator organisms depends on the numbers of their prey. Correspondingly, the number of prey organisms is limited by the number of predators that feed on them. In other words, the size of predator and prey populations are dependent on each other. This relationship depends upon the specific kinds of organisms and the conditions in which they live.

Goals In this investigation, you will **model** interactions between a population of owls and their prey, a population of mice. You will **measure** the sizes of the populations as they change over several generations, and you will **graph** the data you obtain.

LAB WARMUP

Concepts Owls are an example of predator organisms. They feed on smaller organisms such as mice. The mice therefore serve as prey for the owls. As predators, owls occur high in a food chain of forest organisms. Mice occur lower on the food chain. In modeling predator/prey interactions, one needs to make simplifying assumptions. In this investigation, you will assume that the owls feed only on mice. You will also assume that all owls that can catch and eat a certain number of mice will survive and reproduce. These assumptions are like patterns that exist in nature, but do not mirror them exactly. The assumptions are useful, however, in simplifying the model so that population patterns emerge.

Review Section 6.1, Relations in the Ecosystem, should be completed before beginning this investigation. You should also understand the following terms before you perform this investigation.

population   predator   prey   generation   food chain   habitat

Make a **prediction** about the outcome of this experiment and write it in the Lab Notebook.

MATERIALS (PER GROUP)

- metric ruler
- paper
- scissors
- cardboard
- masking tape

© *Addison-Wesley Publishing Company, Inc. All Rights Reserved.*

37

PROCEDURE

1. Use the metric ruler and a sheet of paper to create a grid of squares that are each 1 cm on a side. Draw enough lines to make 200 squares, each of which will represent a mouse. Use scissors to cut out the squares. **CAUTION: Be careful not to cut yourself when using scissors.**

2. Draw and cut out a cardboard square that is 6 cm on a side. This larger square will represent an owl.

3. You will simulate 25 generations of mice and owls that live within a habitat. For this simulation, assume that each mouse not eaten by an owl survives and produces one offspring. To avoid starvation, each owl must catch at least three mice. Assume that each surviving owl produces one offspring for every three mice it has caught. To represent the habitat, place masking tape on the floor or a table top to make a square 30 cm on a side.

4. Place 100 of the mouse squares randomly within the habitat square. Do not allow any to overlap. This set of 100 squares represents the first generation of the mouse population. Set aside the 100 remaining mouse squares in a pile for later use.

5. Hold the owl square at a height of about 30 cm above the habitat square, and drop it onto the habitat. Assume that any mouse square that is at least partially covered by the owl square is a catch. Catch as many of the mice as you can with one drop. Remove and count the captured mice. Assume that there are two owls in the first generation, and drop the owl square a second time to represent the feeding attempt of this second owl. Again, remove and count any mice that have been caught.

6. In column D of the table in the Lab Notebook, record the total number of mice caught by both owls. Note that columns B and C have already been filled in with the original numbers of mice and owls in the first generation.

7. Each mouse that has not been caught is assumed to produce one offspring. Add to the habitat one offspring for each mouse that has not been caught. Use the pile of mouse squares that you reserved earlier as offspring. In column F, record the total number of surviving mice and offspring.

8. Determine whether each of the owls survives and reproduces. (Remember that an owl must catch at least three mice to survive, and that it produces one offspring for every three mice it has caught.) Note and record in column E of the Lab Notebook the number of owls that have starved. Record in column G the total number of surviving owls and their offspring. This number will equal the number of drops to be made by owls in the next generation.

9. Copy the numbers from columns F and G of generation 1 to columns B and C, respectively, of generation 2. These numbers represent the population sizes of mice and owls present at the start of the second generation.

10. Repeat steps 5 through 9 until you have simulated 25 generations. Stop after the owls from the 24th generation feed. Make certain that at the beginning of each generation there are at least three mice and one owl in the habitat. If the populations fall too low, bring the numbers up to these minimum values by adding mouse squares or allowing one owl drop. Also note that the mouse population cannot exceed 200.

11. On a sheet of graph paper, use your data to make a graph of the numbers of mice and owls at the beginning of each generation versus the generation number. Plot the generation number along the *x*-axis. Plot the numbers of mice and owls along the *y*-axis. Use dots to mark the numbers of mice and Xs to mark the numbers of owls. Connect the dots to form a curve for mice. Connect the Xs to form a curve for owls.

Name: _____ Class: _____ Date: _____

LAB NOTEBOOK: INVESTIGATION 8

PREDICTION **A correct prediction is that the numbers of prey and predators will increase at first, then prey numbers will drop off, and predator numbers will drop off correspondingly. The pattern will then repeat, forming a cycle.**

OBSERVATIONS

MOUSE AND OWL POPULATIONS

A	B	C	D	E	F	G
Generation	Number of mice at start of generation	Number of owls at start of generation	Number of mice caught	Number of owls starved	Number of surviving mice and offspring	Number of surviving owls and offspring
1	100	2				
2						
3						
4			Data will vary.			
5						
6						
7						
8						
9						
10						
11						
12						
13						
14						
15						
16						
17						
18						
19						
20						
21						
22						
23						
24						
25						

© Addison-Wesley Publishing Company, Inc. All Rights Reserved.

DATA ANALYSIS

1. What happened to the mouse population during the first few generations? What happened to the owl population during this period?
   The mouse population increased rapidly; the owl population increased, but less rapidly.

2. What happened to the mouse population after many more generations? What happened to the owl population?
   After a certain point, the mouse population dropped dramatically, after which the owl population also fell.

3. Based on your graph, relate the trends in the population sizes of the mice and owls.
   There were initial increases in both the mouse and owl populations. When the owl population reached a certain size, the mouse population decreased rapidly, which then caused a rapid decline in the owl population. This pattern then repeated. Changes in the sizes of the two populations should be related.

CONCLUSION

1. **Infer** Suppose that you were given an unlabeled graph of owl and mouse populations. Given what you observed on the graph you made, how could you infer which curve represented the owls and which represented the mice?
   The prey population is represented by the curve that has larger numbers overall, because there are generally more prey than predators in a community. Also, the major increases and decreases in the predator population follow the corresponding changes in the prey population.

2. **Contrast** Compare your model of interactions between the owl and mouse populations with what might actually occur in a community that includes owls and mice. How do you account for differences?
   Interactions would be more complex. The predators and prey would interact with many more species, and the relative dependency of owl and mouse populations would be less clearly defined. Also, some of the simplifying assumptions made in the simulation (such as that there is one offspring per surviving mouse, and that owls produce one offspring for every three mice they catch) would not hold, further complicating the actual results.

EXTENSION

Predict What would happen if a second animal species that fed on mice was added to the simulation you carried out? Design an experiment to test your prediction. Remember to make your simplifying assumptions before carrying out the simulation.

LABORATORY INVESTIGATION

9 ECOLOGICAL SUCCESSION

Problem: *What evidence of ecological succession is visible in an abandoned field or lot?*

INTRODUCTION

Background   The soil of an area provides the foundation for plant and animal development. Areas that have not yet developed plants tend to have poor, hard, or shallow soil. As plants begin to develop, soil tends to improve.

As soil conditions change, different communities of plants and other organisms may develop over time in a single area. This sequence of communities is called *ecological succession*. Succession generally follows one of two patterns. Succession is said to be *primary* if it occurs in a habitat that initially has no soil or organisms. An example of such a habitat is a lava field after a volcanic eruption. Succession is *secondary* if it occurs as a result of a disturbance. An example of such a disturbance is a forest fire that does not destroy the soil.

Goals   In this investigation, you will **observe** the plants that live in an abandoned field or a vacant lot. You will **compare** the plants and **infer** whether ecological succession is taking place.

LAB WARMUP

Concepts   In a field or lot that has been developed, original communities of plants have usually been cleared away to make room for farms or buildings. When buildings are removed or farms are abandoned, ecological succession often takes place. Various plant species move into the land and replace one another over time. The plant species tend to succeed one another in patterns related to factors such as competition for soil and light. The *pioneer plants*, which develop first, must have the kinds of roots and other adaptations that allow them to live in the harsh conditions that exist early on. The improved soil that results from the growth of these plants then serves as home for plants that could not survive earlier. Eventually, a stable community, called a *climax community*, will develop and remain unchanged unless conditions are disturbed significantly.

A technique called the *transect method* is useful in examining plants within an area. In this method, one or more straight lines are chosen for observation. These lines generally run through regions of the area that contain typical kinds and numbers of plants. The plants that grow along or near the line are studied, and the relative numbers of plant species and their characteristics can be compared.

Review   Section 6.2, Ecological Succession, should be completed before beginning this investigation. You should also understand the following terms before you perform this investigation.

**community   ecological succession   primary succession
secondary succession   climax community   pioneer organism**

Make a **prediction** about the outcome of this experiment and write it in the Lab Notebook.

MATERIALS (PER GROUP)

- 2 thin wooden stakes
- field guide to plants
- thick string, about 6 m long
- trowel
- metric ruler

PROCEDURE

1. Go to a nearby vacant lot or abandoned field selected by your teacher for this purpose. **CAUTION: Go with at least one other student or an adult. Go only to a site that has been approved by your teacher.**

2. Make a survey of the kinds of plants that are common in the area. Choose a part of the area through which you will run your transect. Make sure the transect includes a good cross-section of the plants that grow throughout the area. The transect should be 5 m long and should run toward the middle of the area from one of its edges—for example, from a wooded area or from the wall of a building on the edge of the lot or field.

3. Push wooden stakes into the soil at the beginning and end points of the transect. Connect the two stakes with a string to form a line that is 5 m long.

4. Use a field guide to identify each of the plants you find growing along the transect. Label the drawing of the transect line in the Lab Notebook with the names of the plants in the locations they are found. List each kind of plant each time it occurs. Survey plants that are 1 m or less from the transect line.

5. Examine the various plants more closely. Gently use a trowel to get a sense of whether the roots of each one are shallow, moderately deep, or very deep. Use a metric ruler to determine the typical height of each kind of plant. Record this information in the table in the Lab Notebook.

6. Count the number of each kind of plant along the labeled transect line. On a separate sheet of graph paper, make a bar graph of the numbers of each kind of plant. Plot the kinds of plants along the *x*-axis. Plot the numbers of plants along the *y*-axis.

Name: _____ Class: _____ Date: _____

LAB NOTEBOOK: INVESTIGATION 9

PREDICTION Accurate predictions depend upon the specific plants studied. Generally, succession proceeds from low, shallow-rooted grasses to taller, more deeply rooted plants.

OBSERVATIONS

LOCATIONS OF PLANTS ALONG A TRANSECT

```
|----|----|----|----|----|----|----|----|----|----|
0 m  0.5  1.0  1.5  2.0  2.5  3.0  3.5  4.0  4.5  5.0
```

CHARACTERISTICS OF PLANTS ALONG A TRANSECT

Kind of plant	Typical root depth	Typical height
	Answers will vary.	

DATA ANALYSIS

1. What patterns in the growth of plant species can you observe by examining your transect data?
   Answers will vary, but students should describe general differences in plants growing along the transect, if any difference exists.

2. Which of the plants seem to be pioneers (the first plants to grow) in the area studied? What traits of these plants make you think so?
   Short plants with shallow roots generally are the first to grow.

3. Which plants seem to be following the pioneers and taking over the area in which they grow? What observed characteristics of these plants makes you think so?
   Taller, more deeply rooted plants usually follow pioneers.

CONCLUSION

1. **Interpret data** What is the general order of ecological succession in the field or lot you studied? Which plants do you think will make up the climax community?
   Accept all logical responses.

2. **Classify** Is the ecological succession you observed primary or secondary? Explain your answer.
   Succession was secondary; the disturbance (e.g., clearing, building, farming, and eventual abandonment) disrupted earlier patterns of growth, but did not destroy the soil.

3. **Infer** How do the characteristics of pioneer plant species make them better adapted to conditions in a new community? How do the characteristics of plants that grow later in the area allow them to replace earlier species?
   Short plants with shallow roots that reproduce and spread quickly are well adapted to poor soil conditions that exist early in the succession. Later, as the soil is improved or broken up by the pioneer plants, larger, more deep-rooting plants can grow. These plants compete more effectively for water and light, and gradually crowd out the earlier species.

EXTENSION

Predict What do you think will happen to the relative numbers of the plants in the area you studied over time? To get an idea of whether your prediction is correct, find a nearby lot or field that has similar conditions to the one you studied, but was abandoned earlier. Note whether the plants you thought would come later in succession are more abundant in this field.

LABORATORY INVESTIGATION

10 SCHOOLYARD MICROENVIRONMENTS

Problem: *How do abiotic factors affect the kinds of organisms that live in various microhabitats?*

INTRODUCTION

Background Small areas of the same habitat can be very different. Such areas may contain organisms that are very different. Also, the number of organisms present may vary greatly. These differing areas are called *microhabitats* or *microenvironments*. The physical conditions in such microhabitats also differ. In fact, these conditions are the reason for the differences in kinds and numbers of organisms. Important abiotic factors, such as the amount of light and water, are among the characteristics that define the conditions of each microhabitat. The different abiotic conditions in microenvironments help to determine the organisms found in the the environment.

Goals In this investigation, you will **observe** microhabitats within your schoolyard. You will **compare** the organisms in those microhabitats, and then **control variables** to study the effects of various abiotic factors on those organisms.

LAB WARMUP

Concepts Even within temperate regions, small areas, such as those under eaves and shrubs, may not receive much rainfall. Such areas may be inhospitable to most organisms. Other dry microhabitats include heavily trodden areas, such as paths. Microdeserts can develop in such dry areas. Other small areas may lie near drain pipes or watered lawns. These areas can become microrain forests. The plants and animals that live in these two kinds of microhabitats may differ greatly. Organisms in a microdesert will have adaptations that allow them to live in areas with little water and variable temperature. Organisms in microrain forests will be adapted to high moisture conditions. Abiotic variations in small habitats are similar to the variations in different biomes.

Review Section 7.1, Deserts, should be completed before beginning the investigation. You should also understand the following terms before you perform this investigation.

habitat   microhabitat   desert   rain forest   abiotic factor

Make a **prediction** about the outcome of this experiment and write it in the Lab Notebook.

MATERIALS (PER GROUP)

- metric ruler
- 8 stakes
- 2 5-m lengths of string
- soil thermometer
- trowel
- water
- watch that indicates seconds
- field guide to plants
- field guide to small animals
- opaque (non-see-through) plastic sheeting

PROCEDURE

1. With your teacher's help, observe several small areas within your schoolyard. Look for areas that differ in the amount of moisture, shade, or sunlight they receive. Try to find areas that appear to be a microdesert and a microrain forest.

2. Measure off a 1-m square that includes a microdesert. Place four stakes at the corners. Connect the stakes with string to mark off the square. Do the same for an area that includes a microrain forest.

3. Describe the two locations and the conditions in each microhabitat. Record this information in the first table in the Lab Notebook.

4. Use the soil thermometer to measure soil temperatures at ground level in the morning and in the afternoon. Record your findings.

5. Use the trowel to make a small, shallow circular depression in each area, and pour in equal amounts of water. Note and record the rate at which the water soaks in.

6. Use the field guides to identify and count plants and small animals within each area. **Caution: Do not cause harm or injury to plants and animals in your area.** Note any adaptations that make them suited to life in the microhabitat. You may use a trowel to gently probe the soil in each area. Observe insects or worms within each. Also observe whether the plants are shallow-rooted or deep-rooted. Record this information in the second table in the Lab Notebook.

7. Now vary the conditions within the two microhabitats. For example, you can hang dark plastic sheeting on the strings and stakes. Do this to cover the sunnier area or to reduce rainfall on the wetter area. You can water the ordinarily dry area. Maintain these changed conditions for a period of at least three weeks. Observe and record what happens to the organisms within the areas.

Name: _____ Class: _____ Date: _____

LAB NOTEBOOK: INVESTIGATION 10

PREDICTION **A correct prediction is that organisms living in a certain microhabitat have adaptations that are suited to the abiotic conditions of the microhabitat.**

OBSERVATIONS

	Microdesert	Micro-rain forest
Light conditions		
Soil temperature A.M.	**Observations will vary.**	
Soil temperature P.M.		
Drainage time		
Kinds and numbers of plants		
Kinds and numbers of animals		

Plant adaptations	Animal adaptations
Answers will vary.	

© Addison-Wesley Publishing Company, Inc. All Rights Reserved.

47

DATA ANALYSIS

1. How did the numbers and kinds of organisms in the two microhabitats vary? How would you explain these differences?
   Answers will vary. The microdesert probably contained fewer organisms because less water was available.

2. How did the adaptations of the organisms in the two microhabitats vary? What adaptations, such as root depth, were common to most of the plants within a microhabitat?
   In general, organisms in the microdesert had adaptations (e.g., widely spreading, shallow roots) that enabled them to make the most of limited resources and harsh conditions. Plants in the micro–rain forest were probably leafier, taller, and more deeply rooted.

3. Compare conditions in the two microhabitats after you altered their abiotic factors. How did the plants that did well under the new conditions in one microhabitat compare to those that did well under the original conditions in the other microhabitat?
   Most plants that did well originally probably did poorly under the new conditions. Kinds of plants that were absent or present only in small numbers probably began to spread into the area. Plants that did well under the new conditions were probably similar to those that did well originally in the other microhabitat.

CONCLUSION

1. **Generalize** Summarize the differences in the abiotic factors of the two microhabitats.
   The microdesert received little water and probably a good deal of sunlight. The micro–rain forest probably received a great deal of water and less sunlight.

2. **Infer** How did the adaptations of the organisms you observed benefit the organisms in the microhabitats in which they lived?
   Microdesert plants have spreading, shallow roots to absorb rapidly draining water, and small, thick leaves to minimize evaporation. Microrain forest plants might have deeper roots to absorb water from saturated soil, and broad foliage from abundant water. Animals in each microhabitat would be adapted to the abiotic factors.

EXTENSION

Model How would you design a garden that includes organisms native to very different habitats? Describe the kind of garden you would design. Consider its location in relation to the path of the sun or to large shade-producing plants, and other important features. Describe what special structures or overhangs you might construct, and how you would vary and control abiotic factors. Make a detailed drawing of your proposed multihabitat garden.

LABORATORY INVESTIGATION 11

CHAPARRAL AND FIRE ECOLOGY

Problem: *What factors cause certain kinds of seeds to germinate only after a fire has occurred?*

INTRODUCTION

Background   Chaparral is an ecosystem of short trees and shrubs that forms under conditions of long, hot, dry summers and cool, wet winters. Chaparral forest growth is typical of southern and central California and the Mediterranean coast of Europe. The seeds of chaparral plants and grassland plants do not often germinate except under specific conditions.

Fire plays an important role in maintaining the chaparral community. In the absence of fires, chaparral plants of many species are less successful in developing from seed. When fires do not occur frequently enough, new plants do not grow in large numbers, and the chaparral community becomes less healthy and varied.

Goals   In this investigation, you will **observe** the germination of typical chaparral seeds under varying conditions. You will **control variables** and **interpret data** in order to **infer** which variables are important to the germination of seeds from different species.

LAB WARMUP

Concepts   Fires are common in dry areas such as chaparral and grassland regions. The seeds of many plants growing in such regions germinate only after fires have destroyed mature vegetation around them. A number of variables may account for this effect, including heat from the fire and the presence of certain substances released from charred vegetation. In this investigation, you will study the possible effects, both separately and together, of two of the variables—heat and substances resulting from charring.

Review   Section 8.2, Steppes and Prairies, should be completed before beginning this investigation. You should also understand the following terms before you perform this investigation.

chaparral   germination   control variable

Make a **prediction** about the outcome of this experiment and write it in the Lab Notebook.

MATERIALS (PER GROUP)

Two of the following species of seeds:
- 40 camas (*Camisonia californica*) seeds, 20 of them heat treated
- 40 buckthorn (*Ceanothus megacarpus*) seeds, 20 of them heat treated
- 40 whispering bells (*Emmenanthe penduliflora*) seeds, 20 of them heat treated
- 40 penstemon (*Penstemon spectabilis*) seeds, 20 of them heat treated

- 8 petri dishes with covers
- grease pencil
- fine potting soil
- water
- char extract (water in which charred and ground balsa or birchwood has been soaked)

PROCEDURE

1. Obtain 40 seeds of each of the two species assigned to you by your teacher. Twenty of each kind of seed will have been heat treated, so you will have four sets of seeds: species 1, unheated; species 1, heated; species 2, unheated; and species 2, heated.

2. Obtain eight petri dishes. Label four of them with the common name of one of the species. Add the following four labels to these four dishes: *"no char, no heat," "char only," "heat only," "both char and heat."* **Caution: Use care when handling glassware.**

3. Label the remaining four dishes with the common name of the second species. Then add the same four labels as above to these four dishes. You should now have eight dishes with no two labeled exactly alike, as shown in Figure 11.1.

4. Fill the eight petri dishes half full of fine potting soil.

5. Place a total of ten seeds onto the soil in each of the eight labeled petri dishes. Place the seeds according to their species and treatment. See the labeling on the dishes in the illustration below. Non-heat–treated seeds should be placed in the dishes labeled "no char, no heat" and "char only." Heat–treated seeds should be placed in the dishes labeled "heat only" and "both char and heat."

6. To each of the four petri dishes that are labeled "char only" or "both char and heat," add enough char extract to moisten the soil thoroughly. Use water only to thoroughly moisten the soil in the other four dishes. Make sure that the soil is equally moist in all dishes.

7. Cover all the petri dishes. Store them in an area of moderate light, watering the soil from time to time as necessary. Count the number of seedlings that have germinated in each dish during the next four weeks. Divide the number germinated in each by 10, the total number of seeds in the dish. Record this information in the table in the Lab Notebook.

8. Exchange data with students who studied the two species you did not study. Record their data in the table in the Lab Notebook.

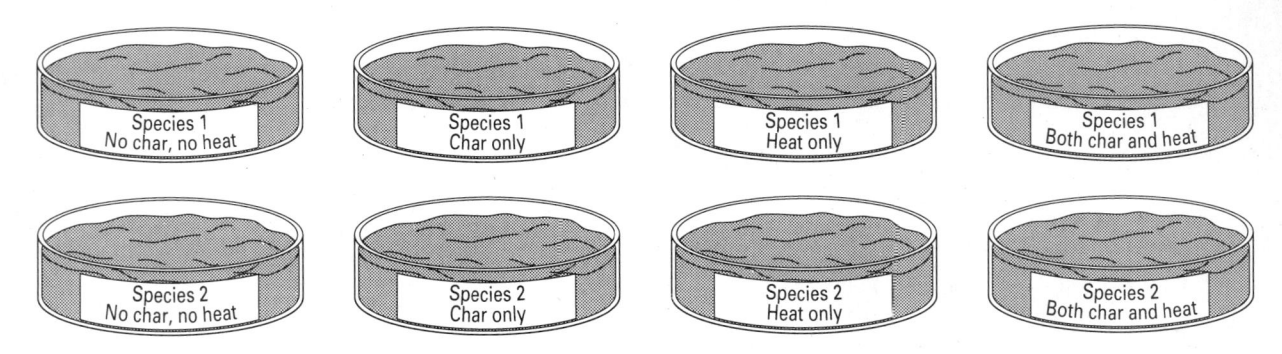

Figure 11.1 Labeled petri dishes

Name: _____ Class: _____ Date: _____

LAB NOTEBOOK: INVESTIGATION 11

PREDICTION **A correct prediction is that both heat and substances resulting from charring promote the germination of chaparral seeds.**

OBSERVATIONS

NUMBERS OF SEEDS GERMINATED

Kind of seed	Treatment of seed	Number germinated	Total
Camas	No char, no heat		
	Char only		
	Heat only		
	Both char and heat		
Buckthorn	No char, no heat		
	Char only		
	Heat only		
	Both char and heat		
Whispering bells	No char, no heat		
	Char only		
	Heat only		
	Both char and heat		
Penstemon	No char, no heat		
	Char only		
	Heat only		
	Both char and heat		

Data should show that seeds that were exposed to heat and/or char germinated in the greatest numbers.

© Addison-Wesley Publishing Company, Inc. All Rights Reserved.

DATA ANALYSIS

1. Which kinds of seeds, if any, germinated in significant numbers with neither heat nor char? What was the purpose of planting some seeds that received neither treatment? What is such a setup called in an experiment?
   None of the seeds was as likely to germinate with neither heat nor char. This setup, called a control, in which neither variable is present, is helpful as a basis of comparison, and helps establish the causal relationship between variables and experimental treatment.

2. Which kinds of seeds, if any, required only heat, but not char, to germinate in significant numbers? Which required only char?
   Buckthorn was able to germinate fairly well with heat alone. Camas, whispering bells, and penstemon required only char.

3. Which kinds of seeds, if any, required both char and heat to germinate maximally?
   Buckthorn required both heat and char to germinate well.

4. On a separate sheet of graph paper, prepare a bar graph of your data. Plot the experimental treatment along the x-axis and the percentage of germination along the y-axis. Does your graph help you to visualize trends in the data?
   The graph does make comparison and visualization of trends easier.

CONCLUSION

1. **Infer** How might char treatment affect seed germination? Why might such an effect be advantageous for seeds growing in chaparral or grassland regions?
   Chemical substances released by burned wood enter the seed and trigger its germination. The frequent chaparral and grassland fires that burn off the competing older vegetation cause these substances to be produced, thereby "signaling" the seeds that conditions are favorable for the growth of new plants.

2. **Infer** How might heat treatment affect seed germination? Why might such an effect be advantageous for seeds growing in chaparral or grassland regions?
   Heat may help crack hard seed coats or may produce chemical changes that induce germination. The fires that destroy competing vegetation produce heat, which "signals" seeds that conditions are favorable for growth.

EXTENSION

Predict In addition to heat and charring, light may also play a role in chaparral and grassland seed germination. Predict whether light or darkness will have a positive effect on the germination of the kinds of seeds you have studied. Then design an experiment to test your prediction. Be sure to test for combined effects as well.

LABORATORY INVESTIGATION

12 FORESTRY AND CONSERVATION STUDY

Problem: *How many board feet of wood are needed to build a new 100-house development?*

INTRODUCTION

Background   To develop a sensible approach to conservation of trees, it is helpful to get an idea of the amount of wood in a tree. It is also important to know the amount harvestable per acre and the size of the area that must be logged to provide the lumber for a specific purpose, such as building a house. The amount of lumber that can be harvested from a tree depends upon the height and diameter of the tree. A useful measure of tree volume is a unit called the *board foot*. One board foot has a volume equal to that of a block of wood that is 12 inches long, 12 inches wide, and 1 inch thick.

Goals   In this investigation, you will first construct a device called a *clinometer*. You will then **measure** the height and circumference of each of the trees in a 1/100-acre wooded area. Next, you will **calculate** the number of board feet of wood in the trees and the total number per acre. Finally, you will **infer** the number of acres that must be cut down to build 100 houses.

LAB WARMUP

Concepts   Two measurements are required to calculate the board feet of lumber in a tree. One measurement is the diameter of the tree at breast height (dbh), which is the diameter 4.5 feet (1.4 m) above the ground. The dbh of a tree is calculated by measuring the circumference of the tree, in inches, and then dividing by 3.14, the approximate value of *pi*.

The second measurement needed to calculate board feet is the tree's height. A clinometer, basically a protractor with a weight on a string, can be used to estimate the height of a tree. With this device, you can determine where to stand so that you are viewing the tree top at an angle of 45° to the ground. The sum of the height of your eyes above ground level and your distance from the base of the tree is about equal to the height of the tree.

Review   Section 9.1, Coniferous Forests, and 9.2, Deciduous Forests, should be completed before beginning the investigation. You should also understand the following terms before you perform this investigation.

board foot   clinometer   diameter   circumference   conservation

Make a **prediction** about the outcome of this experiment and write it in the Lab Notebook.

MATERIALS (PER GROUP)

- flexible tape measure
- paper clip
- string
- protractor
- steel washer
- yardstick
- 4 wooden stakes
- marker

© Addison-Wesley Publishing Company, Inc. All Rights Reserved.

PROCEDURE

1. Use a tape measure to measure the height, to the nearest foot, of your eyes above ground level where you are standing. Record this information in the table in the Lab Notebook.

2. Construct a simple clinometer by using a paper clip to attach a string to the center of the straight edge of a protractor, as shown in Figure 12.1. Tie a steel washer to the free end of the string. The washer should swing freely when the protractor is tilted.

3. Proceed to the wooded site chosen by your teacher. **CAUTION: Go only in the company of your teacher and stay close to your group.** Measure off a 1/100-acre area of the forest by using the yardstick to measure a square of land that is 21 ft by 21 ft (6.5 by 6.5 m). Make sure that the area you select is representative of the forest in terms of the sizes and numbers of trees present. Mark off the corners of the square with stakes.

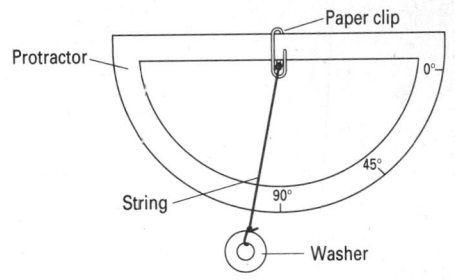

Figure 12.1 Clinometer

4. Working from one side of the square to the opposite side, measure the approximate dbh and height of each of the trees in the 1/100-acre area. (Disregard trees less than about 16 ft [4.9 m] high.) To estimate the dbh of a tree, use the yardstick to measure 4.5 ft (1.4 m) up from the ground. Use the tape measure to determine the circumference of the tree in inches. Repeat this process for all the trees in your area. Record this information the Lab Notebook.

5. Estimate the height of each tree in your area in the following way. Move back from the tree, holding the protractor at eye level with the 0° corner closest to your eye, and sight the top of the tree along the straight edge. **CAUTION: Be careful as you move back. It is advisable to have another student "spot" for you during this process.** Note that when you are close to the tree, the protractor must be greatly tilted to allow you to sight the tree top. The angle then indicated by the string is close to the 0° mark on the protractor. As you continue to move back, however, you must reduce the tilt of the protractor more and more in order to keep sighting the tree top along the straight edge.

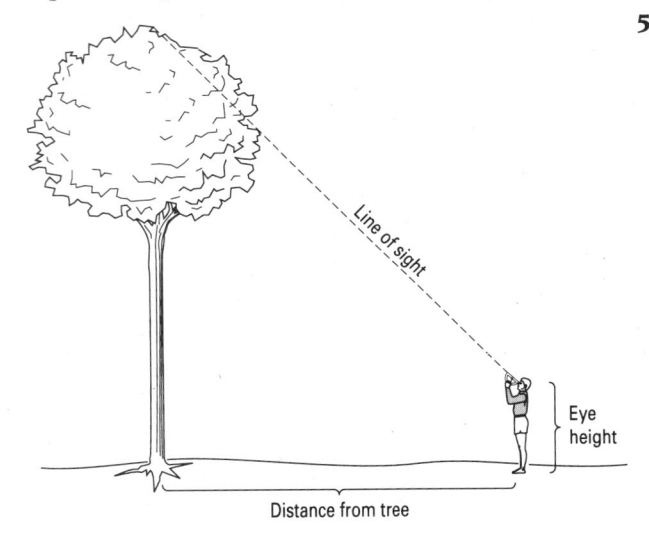

Figure 12.2 Estimating height of a tree

6. When you have moved far enough from the tree so that the protractor string indicates an angle of 45°, stop. Mark your position on the ground, and then use the yardstick to measure your distance from the base of the tree. Record this information in the table. Repeat steps 5 and 6 for each of the trees in your area.

7. On your own, contact a local builder and find out the board footage of wood needed to build a typical house in your area. Record this information in the Lab Notebook.

8. Calculate the approximate height of each tree by adding your eye height to your distance from the tree when you could sight the top at 45°. Record this information in the Lab Notebook.

9. In board-foot calculations, tree heights are generally expressed in terms of numbers of 16-ft lengths. Divide each tree height by 16, round the results to the nearest 0.5, and record this information in the Lab Notebook.

10. Calculate the dbh for each tree. (Divide circumference by 3.14.) Record this information.

© Addison-Wesley Publishing Company, Inc. All Rights Reserved.

Name: _____ Class: _____ Date: _____

LAB NOTEBOOK: INVESTIGATION 12

PREDICTION   **A correct prediction is that a large number of trees must be cut down to build a 100-house development.**

OBSERVATIONS

Height of eyes above ground level: _____

TREES DATA FOR A SELECTED 1/100-ACRE AREA

Tree	Circumference at breast height (in)	Distance from tree (ft)	Height of tree (ft)	Height in 16-ft lengths	Diameter at breast height (in)	Board feet
1						
2						
3			**Data will vary.**			
4						
5						
6						
7						
8						
9						
10						

Use the other side of this sheet if you are measuring more than ten trees.

Number of board feet needed to construct a typical house: _____

BOARD FOOT TABLE

| dbh (in) | Number of 16-ft lengths |||||||
	1.0	1.5	2.0	2.5	3.0	3.5	4.0
14	73	98	123	143	163	175	187
16	76	102	129	150	173	188	204
18	77	104	132	155	178	195	212
20	78	107	136	159	184	202	220
22	80	110	139	164	188	209	228
24	80	110	140	166	193	211	230
26	81	112	143	170	197	217	237

© Addison-Wesley Publishing Company, Inc. All Rights Reserved.

DATA ANALYSIS

1. The table on the previous page gives the approximate number of board feet for trees in general. (Actual numbers differ from species to species.) Use the table to determine the number of board feet in each tree you are measuring. First, find the tree's rounded number of 16-ft lengths (1, 1.5, etc.) across the top of the table. Then find the dbh (14, 16, etc.) closest to that of the tree. The approximate board feet of the tree is at the intersection of the dbh row and the 16-ft-lengths column. Record this information in the Lab Notebook.

2. Calculate the sum of the board feet in all the trees in the 1/100-acre area. Then multiply this number by 100, to obtain the number of board feet in a typical acre of local forest. Record the number and your calculations below.

   Answers will depend upon the particular trees measured.

3. Multiply by 100 the number of board feet needed to construct a single typical house to obtain the number needed to construct a 100-house development. Record the number and your calculations below.

   Answers will depend upon local wood use per typical house.

4. Calculate the number of acres of trees that must be felled to construct the 100-house development. To do so, divide your answer to Data Analysis question 6 by your answer to question 5.

   Answers will vary, depending on students' answers to questions 5 and 6.

CONCLUSION

1. **Interpret data** Given the result of your last calculation, how do major periods of housing construction affect the forests used for lumber?

   The large number of acres of trees felled suggests considerable deforestation during major periods of construction.

2. **Infer** What could be done to help reduce the harvesting of wood for house construction?

   Suggestions include: using other building materials, such as masonry; constructing multiple-unit instead of single-unit housing; reusing old wood; and legislation that more stringently regulates logging.

EXTENSION

Research Estimate how many acres of timber you think are actually harvested commercially each year for all purposes, including export, in the United States. Conduct library research to find out the actual quantity and also the total number of forested acres in the United States that are harvestable in practical terms. Compare the numbers to determine how long it would take, without replanting, to deplete U.S. forests of timber at the current rate of harvesting. Write a paper on your findings.

LABORATORY INVESTIGATION

13 A SURVEY OF PLANKTON COMMUNITIES

Problem: *How diverse are plankton communities in bodies of water with varying levels of pollution?*

INTRODUCTION

Background Plankton are any of a large variety of organisms that drift on or near the surface of water. Most plankton organisms are protists. Plankton that are able to undergo photosynthesis and other plantlike functions are called *phytoplankton*. Many kinds of phytoplankton are algae. Plankton that do not undergo photosynthesis and act as consumers in the ecosystem are called *zooplankton*. Many ocean zooplankton are small larval animals.

Goals In this investigation, you will use a microscope to **observe** plankton in three samples of pond water that differ in their levels of pollution. You will **read a diagram** to identify the kinds of plankton in the samples. Based on the degree of diversity, you will **infer** the relationship between plankton diversity and pollution.

LAB WARMUP

Concepts The health of an ecological community is often reflected in the diversity of the community, or the number of different species it contains. When certain kinds of stresses affect an ecosystem, some organisms sensitive to those stresses may not survive. Community diversity is thereby reduced. This reduction in diversity, in turn, may reduce the stability of the community. This further diminishes its ability to survive and adapt to future stresses.

Review Section 10.2, Standing-Water Ecosystems, should be completed before beginning this investigation. You should also understand the following terms before you perform this investigation.

plankton   phytoplankton   zooplankton   community   ecosystem

Make a **prediction** about the outcome of this experiment and write it in the Lab Notebook.

MATERIALS (PER GROUP)

- samples taken from each of three dropper bottles of pond water labeled *unpolluted, slightly polluted,* and *moderately polluted*
- 3 glass slides
- 3 coverslips
- compound microscope

PROCEDURE

1. **CAUTION: Be careful when handling glass objects and microscopes.** Prepare a wet mount by placing several drops of the unpolluted pond water onto a microscope slide. Put a coverslip over the drop.

2. Observe the wet mount under low to moderate magnification through a compound microscope. Identify in the sample as many of the kinds of plankton as you can. Use the illustration of plankton on page 58 as a guide. Count the number of individuals of each kind. Record this information in the table in the Lab Notebook.

3. Repeat steps 1 and 2, observing samples of the slightly polluted and moderately polluted pond water. In the case of plankton that were previously listed in step 2, simply fill in the numbers of individuals of each kind now observed. Record a zero if a kind of plankton observed earlier is not present. Add the names of any plankton not observed in previous samples.

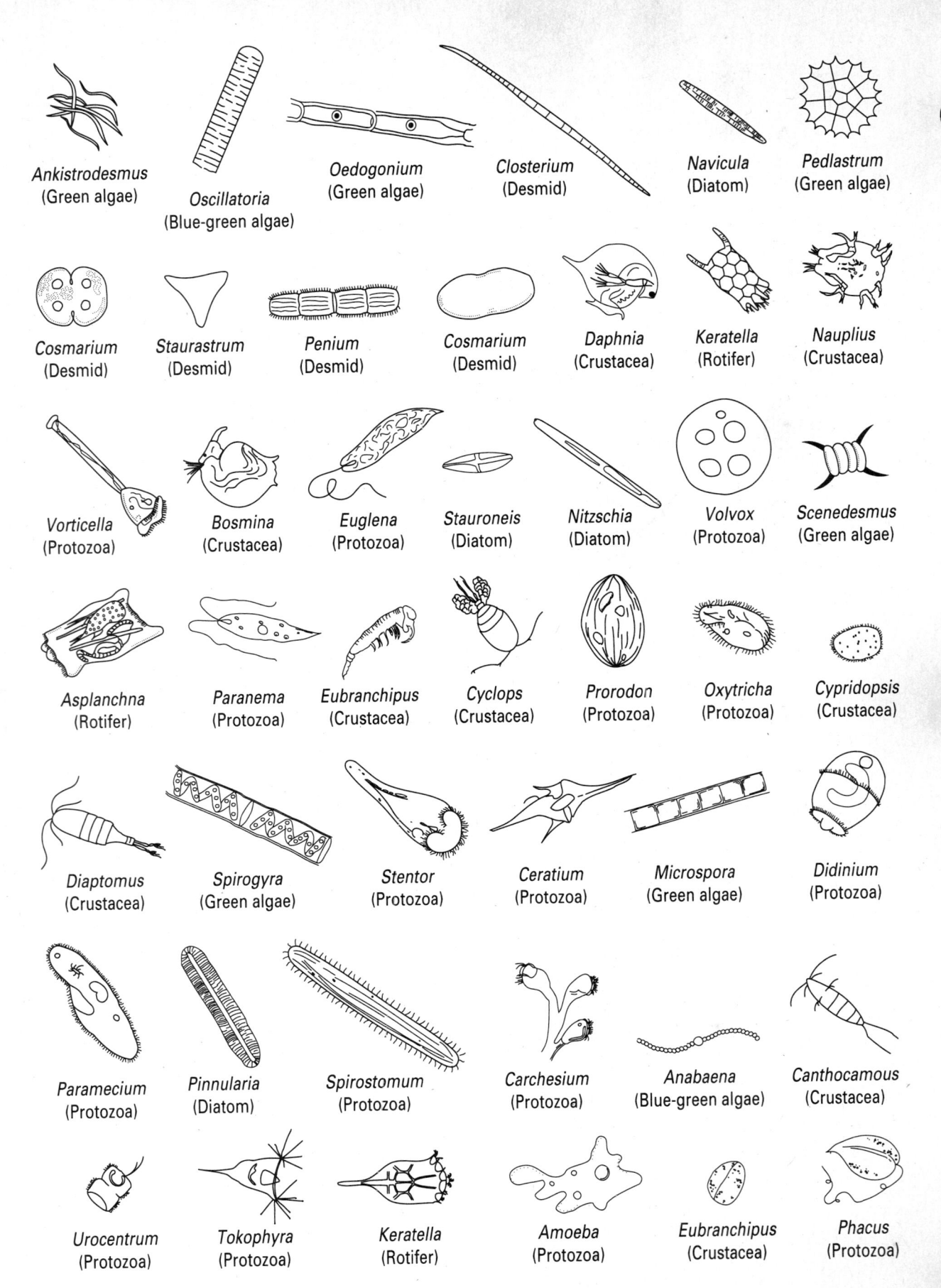

Figure 13.1 Various types of plankton

58

Name: _____ Class: _____ Date: _____

LAB NOTEBOOK: INVESTIGATION 13

PREDICTION **A correct prediction is that the plankton community in the most polluted pond water will be the least diverse.**

OBSERVATIONS

PLANKTON DIVERSITY IN DROPS OF POND WATER FROM THREE SOURCES

Kind of plankton	Number of individuals in source that is:		
	Unpolluted	Slightly polluted	Moderately polluted

Data will vary but should show the greatest diversity of plankton in the unpolluted water.

© Addison-Wesley Publishing Company, Inc. All Rights Reserved.

DATA ANALYSIS

1. Which samples of pond water would you say had the greatest community diversity? The least? Explain your answer.
   The unpolluted source had the greatest diversity because it had the largest number of kinds of organisms. The moderately polluted source had the least diversity.

2. Compare the numbers of individuals of each kind of plankton in the three samples of pond water. Did all kinds of plankton differ in number in the same way from one sample to another?
   For most kinds of plankton, the number of individuals was largest in the unpolluted source and smallest in the moderately polluted source. Different kinds of organisms were affected to different degrees, however.

3. Which sample contained the largest total number of individuals? The smallest number?
   The unpolluted sample contained the most; the moderately polluted sample contained the least.

CONCLUSION

1. **Infer** Based upon your observations, what would you expect to see if you examined highly polluted water?
   There would be very few, if any, kinds of plankton, and the total number of individual organisms would be very low.

2. **Generalize** Make a general statement relating community diversity and pollution. How do you account for this relationship?
   The greater the level of pollution, the lower the community diversity. Pollutants are harmful to most organisms, and may kill off certain populations of organisms entirely, reducing the number of kinds of organisms that make up the community.

3. **Infer** Based upon your observations, how could a study of plankton help identify the kinds of pollutants that are present in a body of water?
   Different kinds of plankton differ in sensitivity and selectiveness in regard to pollution. Comparisons of numbers of specific kinds of plankton may therefore help reveal which pollutants are present and in what concentrations.

EXTENSION

Observe Monitor changes in communities in a local pond by collecting samples every few weeks over a period of several months. Try to account for the changes on the basis of variations in conditions, including the degree of pollution, the season, depth of water in the pond, and temperature.

LABORATORY INVESTIGATION

14 DIATOMS AS WATER-QUALITY INDICATORS

Problem: *How can the diversity of a population of diatoms be measured and used to assess water quality?*

INTRODUCTION

Background *Diatoms* are a large and diverse group of one-celled golden-brown algae. The cell wall of all diatoms is made up mainly of silica and is much like glass in composition. The wall has two parts of slightly different size. The larger half fits just over the smaller one, like the lid on a box. Different species of diatoms are generally easy to distinguish because of differences in the shapes of their shells and in the markings on their cell walls.

Goals In this investigation, you will collect and **observe** a sample of diatoms and **compare** them to tell various species apart. You will then **classify** the sample in terms of its diversity and **infer** the quality of the water from which the sample was taken.

LAB WARMUP

Concepts Diatoms exist in large numbers in almost all naturally occurring water. They are very sensitive to pollution and other factors harmful to water quality. These characteristics make diatoms good indicator organisms in the study of water quality. In general, the smaller the number of kinds of diatoms—in other words, the lower the diversity—the poorer the water quality. Another factor that must also be taken into account is the density. This is the total number of diatoms per unit volume in a sample. A calculation called the *diversity index* (DI) can then be used to determine the health and pollution level of the water ecosystem.

In the case of diatoms, the DI is calculated by first counting the number of times different kinds of diatoms are found next to each other on a microscope slide. The resulting number is then divided by the total number of diatoms observed, to arrive at the DI. The DI scale ranges from 0 to 1.0. A DI that is greater than 0.8 indicates high organism diversity and good water quality. A DI of 0.5 to 0.8 indicates only moderate diversity and suggests some problem with water quality. A DI that is less than 0.5 indicates low diversity and suggests serious problems with water quality.

Review Section 10.3, Flowing-Water Ecosystems, should be completed before beginning this investigation. You should also understand the following terms before you perform this investigation.

diversity   diatom   cell wall   pollution   ecosystem

Make a **prediction** about the outcome of this experiment and write it in the Lab Notebook.

MATERIALS (PER GROUP)

- wooden popsicle stick
- water sample containing diatoms
- collection bottle
- dropper
- coverslip
- mounting medium
- glass slide
- compound microscope

PROCEDURE

1. Go to one of the collection sites suggested by your teacher. **CAUTION: Have an adult accompany you to the collection site. Be careful not to go near areas of deep water.** Scrape some brown-diatom coating from wet rocks, leaves, and twigs, and place it in a collection bottle. Then skim some nearby surface water into the bottle. Record information on the collection site, date of collection, and any other details, in the Lab Notebook.

2. At the beginning of one classroom investigation period, shake the collection bottle to break up any material that has settled to the bottom. Use a dropper to place a drop from the bottle onto a microscope coverslip. Give the slip to your teacher, who will dry it in an oven.

3. During the next class period, obtain the dried, cooled coverslip from your teacher. Put 1 drop of mounting medium onto a glass slide, and place the side of the coverslip with the dried diatom sample onto the medium.

4. Place the slide under a compound microscope set at 40X. Position the slide such that the upper left-hand corner of the coverslip is in the field of vision. Observe the diatom in the upper leftmost position. Place a *d* (for "different") in the first box of the table in the Lab Notebook, to indicate that this is the first of the different kinds of diatoms you observe.

5. Look at the diatom directly to the right of the one you just observed. If it is different from the first one, place a *d* in the second box of the table. If it is the same type, place an *s* (for "same") in the box.

6. Continue observing the diatoms in the row. Each time a diatom is different from the one observed just before, record a *d*. Each time it is the same type as the preceding one, record an *s*. If there are not at least 20 diatoms in the row within the field, move the slide very slightly to view the next field. After you have observed a row of 20, return to the left edge and observe another row. If the first diatom in the row is different from the last in the preceding row, write *d*; otherwise, write *s*.

7. Continue until you have observed 5 rows of 20 diatoms each.

8. In the Lab Notebook, sketch the three most common diatom species.

9. Compare your results with those of a group of students who collected from a different site.

Name: _____ Class: _____ Date: _____

LAB NOTEBOOK: INVESTIGATION 14

PREDICTION **A correct prediction is that water sources that are more polluted or otherwise ecologically disturbed are likely to have a lower diatom diversity than less disturbed ones.**

OBSERVATIONS

Collection site: _____

Description of site and its surroundings: _____
_____
_____

Collection date: _____ Collection time: _____

COMPARISON OF DIATOMS

d = different from preceding diatom; s = same type as preceding diatom

ROW	1	2	3	4	5	6	7	8	9	10	11	12	13	14	15	16	17	18	19	20
1.																				
2.																				
3.																				
4.																				
5.																				

SKETCHES OF MOST COMMON DIATOMS

Diatoms sketched will vary, depending on the individual samples.

DATA ANALYSIS

1. Calculate the diversity index (DI) for your sample by counting the number of *d*'s you recorded in the table and dividing this number by 100, the total number of diatoms you observed. The number you obtain will be a decimal between 0 and 1.
   Answers will vary.

2. Given what you learned about the DI scale in the Concepts section above, how would you rate the diversity of the diatom sample you collected?
   A DI number of more than 0.8 indicates high quality; 0.5 to 0.8 indicates moderate diversity; less than 0.5 indicates low diversity.

3. How would you rate the quality of the water from which you collected the sample?
   A DI number of more than 0.8 indicates good water quality; 0.5 to 0.8 indicates some problem with water quality; less than 0.5 indicates serious problems with water quality.

4. What was the DI value of the sample collected by other groups of students whose results you studied? How would you rate the diatom diversity and water quality of their sample?
   Answers will vary. Diatom diversity and water quality depend on the DI value obtained.

CONCLUSION

1. **Interpret data** Given what you observed directly about your collection site and its surroundings, what environmental factors do you think may have contributed to the level of diatom diversity and water quality?
   Answers will vary, but should reflect the diatom diversity observed and the surrounding conditions of the collection site. For example, the presence of a nearby industrial site, farm fields, or animal feedlots might be cited to account for poor diatom diversity and water quality.

2. **Interpret data** What environmental factors do you think may have contributed to the diatom diversity and water quality of the source from which the other group of students obtained their results?
   Answers will vary but should be based on the DI value calculated by the other group.

3. **Integrate** What changes might be made to increase water quality in the areas studied by you or other students? What difficulties might be encountered in attempting to make such changes?
   Answers will vary. Suggestions might include ideas such as legislating stricter pollution standards. Problems of expense and implementation might be cited as possible difficulties encountered.

EXTENSION

Predict What level of diversity do you predict for other kinds of algae, such as green algae, in the sites you studied? Design and carry out an experiment to test your prediction. Obtain your teacher's permission before proceeding.

LABORATORY INVESTIGATION

15 CLEANING UP OIL SPILLS

Problem: *What effect do detergents used to clean up oil spills have on birds, and what are the most effective ways of cleaning up oil spills?*

INTRODUCTION

Background   Oil spills at sea can be extremely destructive to the environment and the organisms in it. Animals that come in contact with the oil can be poisoned, smothered, or otherwise harmed. Rapid cleanup of oil spills reduces their damaging effects, but no method used alone is very effective. Methods for cleanup currently include the use of skimmers, devices that pass contaminated water through a moving, porous belt that collects the oil for removal. Devices called booms are like large fences placed around a spill to contain it. Oil-absorbing materials, such as hay, are sometimes dropped onto spilled oil, and the resulting oil-saturated debris is then scooped up. Sometimes detergents are used to dissolve the oil. Burning off the oil has also been attempted, though not effectively. The thin layer of oil does not burn readily, and produces air pollution if it burns at all.

Goals   In this investigation, you will observe the effect of a detergent on oil that has been spilled on water, and you will **infer** its effect on bird feathers. You will then **model** a set of techniques that you think will be most effective in cleaning up an oil spill. You will test your techniques by using them to remove oil from a container of water.

LAB WARMUP

Concepts   Some of the methods used to clean up oil spills are directly harmful because they pollute the water, harm wildlife, or even spread the oil. Cleaning oil spills can be harmful to waterbirds because the substances used are harmful to the natural oils that coat the birds' feathers. Cleaning substances can break up or dissolve these oils, which waterproof and insulate birds, providing buoyancy and warmth even in icy water.

Review   Section 11.1, The World Ocean, and Section 21.2, Chemical Pollutants, should be completed before beginning this investigation. You should also understand the following terms before you perform this investigation.

pollution   skimmer   boom   solubility

Make a **prediction** about the outcome of this experiment and write it in the Lab Notebook.

MATERIALS (PER GROUP)

- 2 beakers containing a mixture of water and crude oil or gear lube oil
- tablespoon
- dishwashing liquid
- any other approved materials you decide to use to clean up the oil
- rubber gloves

PROCEDURE

PART A

1. Obtain from your teacher a beaker containing water and a small amount of oil. **CAUTION: The oil is poisonous. Wear rubber gloves, an apron, and goggles. Be careful not to spill any oil.** In the Lab Notebook, describe the appearance of the mixture.
2. Add 1 tablespoon of detergent to the beaker, and stir gently to avoid creating air bubbles. Record your observations.
3. During the following week, devise a plan for removing spilled oil from water quickly and effectively. The plan should take no more than 5 minutes to carry out. You can invent simple devices or use any safe materials you wish. Describe the materials and procedure in the Lab Notebook, and present the plan to your teacher for approval. **Note: Burning the oil will not be allowed.**

PART B

1. The following week, bring to class the materials your teacher approved. Obtain a beaker of water and oil from your teacher. **CAUTION: The oil is poisonous. Wear rubber gloves, an apron, and goggles. Be careful not to spill any oil.**
2. When your teacher says "Go!" begin to carry out your oil-spill removal technique. Your success will be judged on the basis of both speed and effectiveness. When you have finished to your satisfaction, raise your hand and record the amount of time, in minutes, you have taken. Describe the final appearance of the beaker and its contents in your Lab Notebook.
3. Your teacher will evaluate the effectiveness of your oil removal technique and award a score from 1 to 5 (1 for completely effective and 5 for completely ineffective). Record the score your group received.

Name: _____ Class: _____ Date: _____

LAB NOTEBOOK: INVESTIGATION 15

PREDICTION **A correct prediction is that the detergent will cause the oil to break up into droplets. Correctness of predictions regarding cleanup methods depends on the efficiency of the methods devised by students.**

OBSERVATIONS

PART A

Initial appearance of oil-and-water mixture:
The oil forms a layer on top of the water.

Appearance after adding and mixing detergent:
The oil breaks up into droplets; some of the oil may also sink.

Description of proposed cleanup materials and procedure:
Answers will vary.

PART B

Completion time:
Answers for Part B will vary, depending on the effectiveness of the techniques students used.

Appearance after cleanup:

Effectiveness score given by teacher:

© Addison-Wesley Publishing Company, Inc. All Rights Reserved.

DATA ANALYSIS

1. How do you think the detergent produced the change you observed in Part A?
   The detergent molecules clung to both the oil and the water, causing the breakup of the oil surface, mixing, and increased solubility of the oil.

2. Add the amount of time taken to carry out your proposed procedure in Part B and the effectiveness score given to you by your teacher. Record the total number below. (Note that the smaller the number, the better the technique in terms of time and effectiveness factors.)
   Totals will vary.

3. Why are both speed and effectiveness important in cleaning up oil spills in natural ecosystems?
   Speed is important, since a slow technique would permit oil to spread farther before it could be cleaned away or contained. An ineffective technique, whatever its speed, will not remove sufficient amounts of oil.

4. Compare your performance results to those of the other groups of students. How much of a range was there in terms of time and effectiveness?
   Answers will vary, depending on the techniques used by other groups.

CONCLUSION

1. **Infer** Why might the addition of detergent to an oil spill be harmful to waterbirds?
   Detergent dissolves or breaks up the protective oil that occurs naturally on the birds' feathers. This destroys the waterproofing and insulation that allow birds to float and stay warm. The detergent may also be poisonous.

2. **Reevaluate** What would you do to improve your oil cleanup method?
   Accept all logical responses. Students might propose a different technique or the addition of new steps to their original technique.

3. **Infer** What factors make the actual cleanup of oil spills at sea particularly difficult?
   Waves, tides, currents, and winds move the oil about quickly and can make it difficult or dangerous to bring in boats and operate equipment.

EXTENSION

Model Write an explanation of how you might adapt your materials and technique, or those of other students, to create a combined method of cleaning up oil spills at sea. Include diagrams or drawings, and discuss why you think your method may represent an improvement over methods used currently.

LABORATORY INVESTIGATION

16 GARBAGE DISPOSAL

Problem: *What are the best ways to dispose of different kinds of trash?*

INTRODUCTION

Background   Enormous amounts of material are disposed of as trash every day throughout the world. Finding suitable sites in which to place trash has become a major environmental problem. Environmental problems are also posed by trash that contains a mix of both toxic and nontoxic materials. Such trash is often disposed of in landfills. These are special sites in which garbage is spread out, packed down, and layered with soil. A thick covering of soil is placed on top. Landfills are not really appropriate for many kinds of trash. They also take up a great deal of valuable space, are unsightly, and can release pollutants into the environment.

Goals   In this investigation, you will **compare** various items of trash in terms of biodegradability, toxicity, and other factors. You will also **infer** which disposal options are most suitable for each item, and **classify** the items accordingly.

LAB WARMUP

Concepts   Trash items are generally disposed in any of four ways: recycling of the material that makes up the item; reuse of the item itself; composting; and disposal in a landfill. Items that are appropriate for recycling are generally made of a single material, such as metal, glass, paper, cardboard, or plastic. Items suitable for reuse may have value if they are in good condition, and can be reclaimed for their original purpose or some other purpose. Vegetable matter is suitable for composting. The landfill option, which is the least desirable, should be reserved for items that do not fit into any of the above categories. Landfill items can be further divided into three categories: biodegradable (can be broken down naturally into simple, harmless substances); nonbiodegradable nontoxic (not readily broken down naturally, but containing or producing no poisonous materials or by-products); and nonbiodegradable toxic (not readily broken down naturally and containing or producing poisonous materials).

Review   Section 12.3, Sustainable Development, and Section 19.1, Solid Wastes, should be completed before beginning this investigation. You should also understand the following terms before you perform this investigation.

biodegradable   nonbiodegradable   toxic   recycle   compost   landfill

Make a **prediction** about the outcome of this experiment and write it in the Lab Notebook.

MATERIALS (PER GROUP)

labeled paper slips prepared by your teacher

PROCEDURE

1. Your teacher will give you a set of paper slips, each of which lists the name of an item of trash. Your teacher will also give you a set of four slips that list disposal options: recycle, reuse, compost, and landfill. Arrange the latter four slips horizontally on a table or desk, like column heads in a chart, leaving enough room below each to place the other slips.

2. Consider the characteristics of each trash item listed on the slips. Sort each of these slips into the appropriate disposal category, placing below the corresponding disposal option slip. You will have 15 minutes to carry out this process.

3. Copy the result of the sorting process onto the table in the Lab Notebook.

4. Your teacher will review your list and place a mark next to each item that has not been correctly sorted.

5. Remember to recycle the paper slips with which you have worked in this investigation.

Name: _____ Class: _____ Date: _____

LAB NOTEBOOK: INVESTIGATION 16

PREDICTION A correct prediction is that most trash items can be recycled, reused, or composted.

OBSERVATIONS

SORTING OF TRASH ITEMS

Recycle	Reuse	Compost	Landfill

Correct sorting will depend on the actual items listed on the slips of paper. Sample recyclable items include newsprint, brown glass bottle, cereal box, aluminum can, broken air conditioner (some parts), colorless glass jar, plastic detergent container. Reusable items include rubber band, slightly worn clothing. Compostable items include coffee grounds, eggshells, vegetable scraps. Landfill items include can containing insecticide, used cat box litter, spoiled beef, can containing paint, styrofoam cup.

© Addison-Wesley Publishing Company, Inc. All Rights Reserved.

71

DATA ANALYSIS

1. After your teacher has reviewed your lists, reassess the placement of each item that you have listed inappropriately. Record this revised information below.
   Answers will vary.

2. What fraction of the items on the lists could be recycled, reused, or composted instead of being dumped in a landfill?
   Answers will vary depending on the lists, but a large fraction of items should be disposed of by options other than landfill dumping.

3. Further classify the landfill items. Which are biodegradable and which are nonbiodegradable?
   Answers will vary, depending on the lists. A typical biodegradable item might be spoiled beef. A typical nonbiodegradable item might be a styrofoam cup.

4. Of the nonbiodegradable landfill items, which are nontoxic and which are toxic? Should items in these two categories be disposed of in the same way?
   Answers will vary, depending on the lists. A typical nontoxic nonbiodegradable item might be a styrofoam cup. A typical toxic nonbiodegradable item might be insecticide. The toxic materials should not be placed in a typical landfill, but should be properly stored and/or treated.

CONCLUSION

1. **Infer** Why is it important to reduce the quantity of garbage that is disposed in landfills? Why is it especially important to reduce the quantity of nonbiodegradable garbage?
   Space for storing garbage is becoming more and more limited as population size increases. Also, some material in garbage presents specific problems, such as toxicity. Resources are also limited, so recycling is crucial in the long run.

2. **Reevaluate** How could your family reduce the quantity of garbage in landfills? How could you alter your habits of spending, consumption, and usage to produce less trash?
   Answers will vary, but may include resolving to reuse, compost, or recycle items and purchasing recycled and biodegradable items

EXTENSION

Predict How does the garbage actually produced by your family in a week fit into each of the disposal categories you have learned about? Examine your household garbage for a week to test your prediction. **CAUTION: Do not touch any hazardous items. Wash your hands after handling the garbage.** Devise a specific plan to reduce your family's garbage production.

LABORATORY INVESTIGATION

17 HUMAN POPULATION GROWTH

Problem: *How does human population growth depend upon family size?*

INTRODUCTION

Background The human population was relatively small until only a few centuries ago. Then, as breakthroughs in medicine, agriculture, and technology were made, the death rate fell markedly and the population began to rise rapidly. At present, the human population is growing exponentially, and is approaching 6 billion. Such growth is causing serious problems in resource distribution, health, waste disposal, and living space.

Goals In this investigation, you will examine and **graph** human population data. You will then **estimate** population sizes that would result given different average family sizes. You will also **infer** consequences of unlimited population growth.

LAB WARMUP

Concepts The following mathematical equation can be used to project changes in populations that are increasing at an exponential rate, such as the human population.

$$N(t) = N(0)e^{rt}$$

$N(t)$ is the population size after the passage of t units of time. The letter t stands for the number of generations, where one generation is often assumed to be about 25 years long. $N(0)$ is the initial population size, r is the exponential growth rate, and e is a natural logarithmic constant that has a value of about 2.72. The quantity e^r is the factor by which the population increases per unit of time. It is given the symbol λ (lambda). The value of λ is approximately equal to the average number of offspring produced per parent. For example, if each pair of parents produces 2 offspring, the number produced per parent would be 2/2, or 1, giving λ a value of 1. If each pair of parents produces 3 offspring, the number produced per parent would be 3/2, for a λ value of 1.5.

Substitute λ for e^r in the equation above to use the following simpler equation:

$$N(t) = N(0)\lambda^t$$

As an example of a calculation using this equation, consider a population that initially includes 100 individuals, 50 males and 50 females. $N(0)$ is equal to 100. What will the population size be when the fifth generation is born (when $t = 5$)? In other words, you wish to know $N(5)$, the population size in the fifth generation. You are given the additional information that each pair of parents produces 3 offspring, all of which go on to reproduce at the same rate. Therefore, λ has an approximate value of 3/2 = 1.5. You can now substitute numbers into the equation, as follows, and solve for $N(5)$:

$$N(5) = 100 \times 1.5^5 = 759$$

Thus the population size at the time of the fifth generation will be about 759.

Review Section 13.3, Challenges of Overpopulation, should be completed before beginning this investigation. You should also understand the following terms before you perform this investigation.

population   growth rate   exponential

Make a **prediction** about the outcome of this experiment and write it in the Lab Notebook.

MATERIALS (PER GROUP)

- computer (if available)
- graphing program

PROCEDURE

1. Study the table below, which gives world human population figures over the past few hundred years.

Date (A.D.)	Population Size
1550	500 million
1700	700 million
1850	1 billion
1900	1.4 billion
1925	1.8 billion
1950	2.5 billion
1975	4 billion
2000	6 billion

2. Use the grid in the Lab Notebook to plot the information in the table above. The graph plots population versus year. If you have access to a computer, you may use it to help you construct the graph.

Name: _____ Class: _____ Date: _____

LAB NOTEBOOK: INVESTIGATION 17

PREDICTION __A correct prediction is that the larger the family size, the greater the growth of the population.__

OBSERVATIONS

GRAPHS OF WORLD POPULATION GROWTH

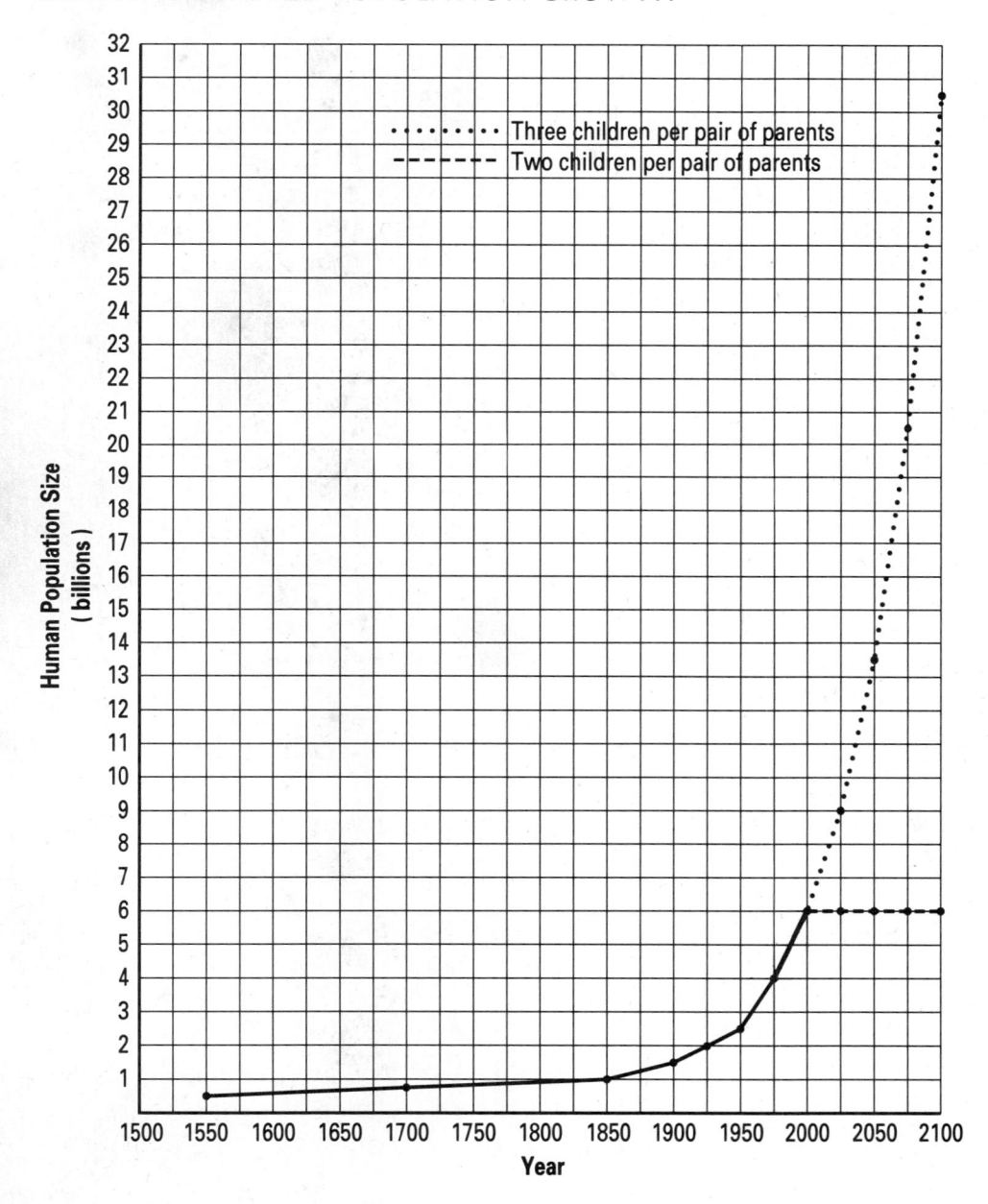

Graph showing Human Population Size (billions) vs. Year (1500–2100). Two curves are plotted: a dotted line representing "Three children per pair of parents" which rises steeply after 2000 to about 30 billion by 2100, and a dashed line representing "Two children per pair of parents" which levels off at about 6 billion after 2000. A solid curve shows historical population growth from ~0.5 billion in 1550 rising to ~6 billion by 2000.

DATA ANALYSIS

1. Calculate λ, given the population sizes in 1975 and 2000.

   λ = 6 billion/4 billion = 1.5

2. Using 6 billion as the value of N(0) in the year 2000, estimate the size of the human population in the years 2025, 2050, 2075, and 2100. Assume that each generation is 25 years long. Also assume that λ has the same value as that calculated in question 2; in other words, growth will continue at the same rate as in the recent past. Add these data points to your graph. Connect them smoothly to the original curve, using a dotted line.

   2025: $N(1) = N(0)\lambda^1$ = 6 billion × 1.5^1 = 9 billion

   2050: $N(2) = N(0)\lambda^2$ = 6 billion × 1.5^2 = 13.5 billion

   2075: $N(3) = N(0)\lambda^3$ = 6 billion × 1.5^3 = 20.25 billion

   2100: $N(4) = N(0)\lambda^4$ = 6 billion × 1.5^4 = 30.375 billion

3. Repeat the calculations you made in question 3, but this time assume that each pair of parents produces only 2 offspring (λ is approximately equal to 2/2, or 1). Add these data points to your graph. Connect them smoothly to the original curve using a dashed line.

   2025: $N(1) = N(0)\lambda^1$ = 6 billion × 1^1 = 6 billion   **2075: $N(3) = N(0)\lambda^3$ = 6 billion × 1^3 = 6 billion**

   2050: $N(2) = N(0)\lambda^2$ = 6 billion × 1^2 = 6 billion   **2100: $N(4) = N(0)\lambda^4$ = 6 billion × 1^4 = 6 billion**

CONCLUSION

1. **Infer** What sustainability problems are created by rapid population growth?

   Accept all logical responses. Living space would become a problem, as would the availability of food, fuel, and other resources. Overall quality of life would likely deteriorate.

2. **Predict** What do you think would ultimately happen if the human population continued to grow at the current rate?

   The survival needs of the population would exceed the capacity of Earth and of technology to meet those needs. The population would be unable to grow beyond a certain point and might decrease catastrophically as a result of famines, overcrowding, etc.

EXTENSION

Graph The average person in the United States uses 280 units of energy per year, compared to 175 units per person in Japan. Use reference materials to find the population sizes in the two countries. Then construct a graph illustrating the differences in energy consumption in each, first assuming 2 offspring per pair of parents and then assuming 3 offspring.

LABORATORY INVESTIGATION

18 DETECTING MUTANT BACTERIA

Problem: *How does an antibiotic affect bacterial growth? How can spontaneous mutations in bacteria resulting in resistance to an antibiotic be detected?*

INTRODUCTION

Background   Some mutations can make an organism resistant to chemicals that originally were harmful to the organism. Bacteria can mutate in this way. Some mutations in bacteria cause a resistance to a particular antibiotic. If the bacteria receive repeated applications of the antibiotic, those bacteria having the resistance will be selected for and will reproduce. A different antibiotic will be needed to control the new bacterial growth. In medicine, this poses a problem when developing and applying antibiotics to treat infections. In agriculture, similar mutations in insects pose a problem in developing and using pesticides.

Goals   In this investigation, you will **observe** the effects of the antibiotic streptomycin on the growth of a bacterial species. You will also **estimate** the *minimum inhibitory concentration (MIC)* of streptomycin necessary to prevent bacterial growth. You will then **isolate** mutant bacteria resistant to streptomycin, and test their resistance to varying concentrations of streptomycin.

LAB WARMUP

Concepts   The organism you will be working with is *Escherichia coli* (*E. coli*), a bacterium common to the intestines of mammals that aids in the digestion of food. You will grow the bacteria on an agar medium that includes proteins, carbohydrates, and other nutrients. *E. coli* is ideal for studying mutations. It reproduces rapidly (about one division every 20 minutes), is easy to maintain, and is relatively harmless. Streptomycin is an antibiotic that interferes with the synthesis of proteins in *E. coli*. Thus, if the agar medium contains streptomycin, bacterial colonies will be unable to form. Some mutations that spontaneously occur in bacteria enable the bacteria to grow in the presence of streptomycin.

Review   Section 14.4, Sustainable Agriculture, should be completed before beginning this investigation. You should also understand the following terms before you perform this investigation.

mutant   mutation   antibiotic   *E. coli*   streptomycin

Make a **prediction** about the outcome of this experiment and write it in the Lab Notebook.

MATERIALS (PER GROUP)

- petri dish with complete agar medium
- petri dish with complete agar medium + 200µg/mL streptomycin
- marker
- inoculating loop
- Bunsen burner
- matches
- *E. coli* in glass test tube with nutrient broth
- 5 plastic bags
- incubator
- metric ruler

- 3 gradient dishes with different maximum concentrations of streptomycin:
    0-20µg/mL
    0-200µg/mL
    0-1000 µg/mL
- glass test tube with nutrient broth
- petri dish with complete agar medium + 20 µg/mL streptomycin
- petri dish with complete agar medium + 200 µg/mL streptomycin
- petri dish with complete agar medium + 1000 µg/mL streptomycin

© Addison-Wesley Publishing Company, Inc. All Rights Reserved.

PROCEDURE

CAUTION: Use care when handling bacterial cultures. Dispose of contaminated supplies only as instructed by your teacher. Wash your hands when you are finished working with the bacteria.

PART A   MINIMUM INHIBITORY CONCENTRATION (MIC) OF STREPTOMYCIN ON <u>E. COLI</u>

1. Obtain the *E. coli* culture broth and three gradient dishes from your teacher, labeled 0-2 µg/mL, 0-200 µg/mL, and 0-1000 µg/mL respectively. These numbers indicate the range of concentrations of streptomycin present in the agar. For example, the streptomycin concentration on one of the dishes is 0 µg/mL on one end (as labeled by your teacher) and gradually increases across the dish to 200 µg/mL on the other end.

2. Mark the bottom of the dishes with five straight parallel lines from the point of zero concentration to the maximum concentration.

3. Inoculate the dishes with *E. coli*. First sterilize the inoculating loop by holding it over the flame of a Bunsen burner for several seconds. Let the loop cool for about a minute, being careful not to touch the loop to anything that might contaminate it. Dip the loop into the broth containing the *E. coli*. Obtain just enough broth to fill the loop—a small amount will contain enough bacteria to form colonies on the dish. Gently drag the inoculating loop across the agar, following the concentration gradient lines drawn on the bottom of the dish. Be careful not to gouge the agar with the loop; barely touch the agar as you drag the loop. Each member of your group should have a turn spreading the bacteria along a different line on each dish. Be careful not to run into other students' bacteria.

4. Place the dishes in plastic bags, close loosely, and invert them. Incubate the dishes at 35°-37°C for 2 to 5 days, until colonies appear.

5. Observe the bacterial growth. The bacteria will grow along the gradient lines up to a point, then you will see no growth. This point is called the minimum inhibitory concentration (MIC). If there is growth all along the gradient, the bacteria in the culture either have an MIC value greater than the maximum concentration of the gradient on the dish, or the bacteria are resistant to streptomycin. If the bacterial growth stops, but you observe one or more colonies growing beyond the MIC point, this indicates that mutant bacteria resistant to streptomycin have grown at those points.

6. Estimate the MIC for each sample using a metric ruler. Mark off intervals $\frac{1}{8}$ the length of the diameter of the disk. Calculate and label the estimated concentration of streptomycin at each mark. For example, the point halfway between 0 and 200 µg/mL has a concentration of 100 µg/mL. Use this as a gauge to estimate MIC values for each line, recording the values in your Lab Notebook. Determine the average MIC value for each dish and record it as well. Draw your observation in the circles provided.

PART B   TESTING STREPTOMYCIN RESISTANCE OF MUTANT BACTERIA

1. Obtain a test tube filled with nutrient broth from your teacher, and label it with your group name, date, and "mutant *E. coli*."

2. With a heat-sterilized inoculating loop, select one mutant bacterial colony from a gradient dish, and transfer some bacteria into the test tube. Mix gently.

3. Incubate the bacterial culture at 35°-37°C until the broth becomes turbid (cloudy). This should take between 24 and 48 hours.

4. Obtain three dishes with concentrations of 20 µg/mL, 200 µg/mL, and 1000 µg/mL of streptomycin, respectively. Inoculate each dish with mutant bacteria from your culture, spreading the bacteria over the agar in a zig-zag manner.

5. Place the dishes in plastic bags, close loosely, and invert them. Incubate the dishes at 35°-37°C for 2 to 5 days, until colonies appear.

6. Count the colonies on each dish, recording your observations in the Lab Notebook.

7. Dispose of all materials as instructed by your teacher.

Name: _____ Class: _____ Date: _____

LAB NOTEBOOK: INVESTIGATION 18

PREDICTION A correct prediction is that streptomycin will prevent bacterial growth. Some mutant bacteria will be resistant to streptomycin.

OBSERVATIONS

PART A

Gradient plate	Line 1 MIC	Line 2 MIC	Line 3 MIC	Line 4 MIC	Line 5 MIC	Average MIC
0–20 µg/mL streptomycin		Data will vary.				
0–200 µg/mL streptomycin						
0–1000 µg/mL streptomycin						

0 µg/mL ○ 20 µg/mL

0 µg/mL ○ 200 µg/mL

0 µg/mL ○ 1000 µg/mL

Observations will vary.

PART B

Number of mutant bacterial colonies

20 µg/mL streptomycin **Data will vary.** _____

200 µg/mL streptomycin _____

1000 µg/mL streptomycin _____

DATA ANALYSIS

1. Why must the inoculating loop be sterilized before inoculating the dishes?
   If the loops were not sterilized, bacteria other than *E. coli* or fungi might contaminate the dishes, giving erroneous results.

2. Did the estimates of MIC vary greatly for your group? Is this what you expected?
   Answers will vary, but since the inoculated bacteria came from the same stock culture, the MIC estimates should be similar.

3. What does bacterial growth beyond the MIC indicate?
   These bacteria are mutants that have a greater resistance to streptomycin.

4. How resistant to streptomycin were the mutant bacteria isolated in Part B?
   Answers will vary, depending on the amount of bacterial growth observed on the three streptomycin dishes.

CONCLUSION

1. **Infer** What is the effect of antibiotics, such as streptomycin, on bacterial growth and the evolution of certain strains of bacteria?
   Antibiotics interfere with the normal functioning of bacteria, preventing their growth. Prolonged exposure to an antibiotic will select for mutant bacteria that are resistant to the antibiot-

2. **Infer** Did streptomycin cause mutations in the bacteria, that is, can streptomycin be considered a mutagen? Explain.
   No; the streptomycin simply set up selective pressures such that only those bacteria resistant to streptomycin would survive to reproduce.

3. **Generalize** This experiment uses a very simple organism to demonstrate resistance to something originally intended to kill the organism. Could a more complex organism such as an insect become resistant to pesticides? Explain.
   A spontaneous mutation in an insect may make it resistant to the pesticide. This resistance will be passed on to its offspring. Since the pesticide will eliminate those insects not resistant to the chemical, the surviving mutant insects will have less competition for food, and thus will proliferate. New strains of insects resistant to the pesticide will emerge.

EXTENSION

Hypothesize What kinds of bacteria are present on the leaves of plants? How does streptomycin affect other species of bacteria? Place a leaf directly on a petri dish filled with complete media and press so that the entire leaf touches the agar. Remove the leaf after 5 or 10 minutes, and incubate. Make sure you do not contaminate the leaf with bacteria from your fingers or some other outside source. Transfer different colonies to dishes with and without streptomycin.

LABORATORY INVESTIGATION

19 BIOLOGICAL PEST CONTROL

Problem: *Can harmful insects be controlled without the use of chemical pesticides?*

INTRODUCTION

Background   One challenge of agriculture is controlling the invasion of insect pests. Many plants of the same species growing closely together attract insects that feed on that species of plant. In natural ecosystems, there is a balance among the population sizes of plants, insect pests, and predators that feed on the insects. In an agricultural setting where predators may not keep the insect population in check, farmers have had to resort to the use of harmful chemicals, called pesticides, to kill insects.

The use of chemical pesticides, in turn, poses a variety of problems. Pesticides kill not only the pests, but also beneficial insects such as natural predators of the pests. When a pesticide enters a food chain, the chemical can become more concentrated in each trophic level. High concentrations of the chemical can seriously threaten the health of organisms in the highest trophic levels. Additionally, some insect species develop a resistance to the pesticides used to kill them. The continued use of these pesticides selects for these more resistant insect strains, making it necessary to develop new and more deadly pesticides.

Biological pest control is an alternative to the use of pesticides. Insects that feed on crops have natural predators. Releasing these predators into the crop area can reduce the pest population, enhancing crop yield without pesticides.

Goals   In this investigation, you will grow cabbage plants and **observe** the effect of aphids on plant biomass. You will also **observe** the effects on aphids and plant biomass when midges, predators of aphids, are added to the plants. You will then **evaluate** the biological control of aphid populations on cabbage plants.

LAB WARMUP

Concepts   The turnip aphid (*Hyadaphis pseudobrassicae*) is a small, green herbivorous insect that feeds on cabbages. The aphid has a proboscis used for feeding. The aphid inserts its proboscis into a cabbage leaf to draw out the sugars in the leaf. Under ideal conditions when food is plentiful, the aphid female can reproduce asexually. In this process a clone of the female parent is produced; no male is needed. However, if food is not plentiful, aphids reproduce sexually (with male and female parents).

The midge (*Aphidoletes aphidimyza*) feeds solely on aphids. The adult females lay their eggs on leaves that are infested with aphids. When the eggs hatch, the larvae find the aphids and inject a toxin into them, killing them, and then eat them. If aphids are very abundant, the midge larvae will kill more aphids than they can eat. After growing to full size, the larvae form cocoons for their pupal stage (lasting 10 to 14 days), after which time adults emerge to continue the cycle.

Review   Section 14.4, Sustainable Agriculture, should be completed before beginning this investigation. You should also understand the following terms before you perform this investigation.

biomass   predation   biological pest control   trophic level

Make a **prediction** about the outcome of this experiment and write it in the Lab Notebook.

MATERIALS (PER CLASS)

- 75 cabbage plants (*Brassica oleracea capitata*) in 3-inch pots
- 75 plastic tabs for labeling plants
- permanent markers
- 3 screened cages
- 500 turnip aphids (*Hyadaphis pseudobrassicae*)
- small paintbrushes
- 250 midge pupae (*Aphidoletes aphidimyza*)
- scissors
- jars with lids (2 per student)
- balances
- water
- 1-L beaker

PROCEDURE

1. Each student will be responsible for two or three potted cabbage plants. Your teacher will assign you pot numbers (A1 through A25, B1 through B25, and C1 through C25). Obtain two or three cabbage plants and plastic tabs, and label them with their assigned numbers and your initials.

2. Place pots A1 through A25 in the screened cage labeled A. Check the leaves of each plant and remove any aphids that may already be there. Do this before any aphids or midges are handled to prevent infestation of the plants.

3. Carefully collect ten aphids one at a time with a paintbrush (to prevent crushing the tiny insects) and place them on cabbage leaves of each remaining plant (B1 through B25 and C1 through C25).

4. Place pots B1 through B25 in the screened cage labeled B.

5. For the remaining pots (C1 through C25), carefully place ten midge pupae in each pot. If the stock of pupae are set on cotton, simply cut around ten midges and place the cotton with the attached pupae directly on the soil.

6. Place pots C1 through C25 in the screened cage labeled C.

7. Allow the plants to grow for 2 to 3 weeks, watering the plants as needed. Be sure to water all plants at the same time and with the same amount of water. Measure the water in the 1-L beaker.

8. After the growing period, bring the plants inside the laboratory for observation, beginning with the pots from cage A.

9. For each assigned plant, cut each leaf off the plant at its base. Checking the entire leaf for insects, place midges in one jar and aphids in another. Repeat this for all other leaves and stems, saving the plant pieces for later analysis. Count the total number of midges and the total number of aphids found on that plant, and record these numbers in the appropriate spaces of the data table in the Lab Notebook.

10. For the same plant, weigh the plant's leaves and stems, recording its biomass in the appropriate space of the data table.

11. Repeat steps 9 and 10 for your remaining assigned plants.

12. Record your data in the class data table on the chalkboard. Record other students' data in your data table in the Lab Notebook.

13. Average the number of aphids per plant for all plants in each cage. Record this average in the Lab Notebook.

14. Average the number of midge larvae per plant for all plants in each cage. Record this average in the Lab Notebook.

15. Average the plant biomass for all plants in each cage. Record this average in the Lab Notebook.

16. **CAUTION: Wash your hands after handling the plants and insects.** Dispose of the plants and insects as instructed by your teacher.

Name: _____ Class: _____ Date: _____

LAB NOTEBOOK: INVESTIGATION 19

PREDICTION   A correct prediction is that the aphids will decrease the biomass of the cabbage plants. The addition of midges will effectively reduce the number of aphids on the plants, leaving more plant biomass intact.

OBSERVATIONS

CABBAGE PLANTS

	A			B			C		
	Number of aphids	Number of midges	Plant biomass	Number of aphids	Number of midges	Plant biomass	Number of aphids	Number of midges	Plant biomass
1									
2									
3									
4									
5									
6				Data will vary.					
7									
8									
9									
10									
11									
12									
13									
14									
15									
16									
17									
18									
19									
20									
21									
22									
23									
24									
25									

CLASS AVERAGES

	Number of aphids per plant	Number of midges per plant	Plant biomass (g)
Cage A	Data will vary, but cage B should have the most aphids and the smallest plant biomass.		
Cage B			
Cage C			

© Addison-Wesley Publishing Company, Inc. All Rights Reserved.

DATA ANALYSIS

1. What is the purpose of the plants in cage A?
   These serve as the control of the experiment.

2. Which cage had the highest average number of aphids per plant?
   The plants in cage B should have the highest aphid numbers.

3. Which cage had the lowest average number of aphids per plant?
   The plants in cage A should have the lowest aphid numbers.

4. Explain how aphids might have gotten onto the plants in cage A.
   Aphids might have come through cracks in the cage or might have entered the cage when it was opened to water the plants.

5. Which cage had the higher average plant biomass, cage B or cage C? Explain.
   Cage C should have the higher biomass, since midges were added to keep the aphid population low, thereby protecting the plants.

6. What is the relationship between aphid population size and plant biomass?
   As the number of aphids increases, the biomass decreases.

CONCLUSION

1. **Infer** Were your results similar to what you predicted? If not, explain what factors may have affected your results.
   Results should correspond to students' predictions. Accidental aphid infestation of the control group, or failure of plants to grow are two problems students may have encountered.

2. **Infer** Is the addition of midges an effective way of controlling aphid populations on cabbage plants? Justify your answer based on your results from this experiment.
   Results should support the conclusion that adding midges effectively reduces aphid populations, leaving more of the plant biomass intact.

EXTENSION

Predict Assume a large field growing only one species of crop has a pest problem, and the farmer decides to use pesticides to kill the crop-eating insects. Predict the short-term and long-term effects of their use.

LABORATORY INVESTIGATION

20 OIL EXTRACTION

Problem: *How can oil be extracted most efficiently?*

INTRODUCTION

Background Petroleum is a heavy, organic liquid formed from the remains of marine organisms that lived millions of years ago. At that time, the sea covered much of the area that is now dry land. When organisms in the sea died, their remains sank to the bottom and mixed with sand there to form layers of sediment. These layers were buried more and more deeply and eventually turned to rock. The pressure of the material on top of the decomposing organic matter helped turn it to petroleum.

Petroleum is a fossil fuel, and is considered nonrenewable because it takes millions of years to form. Also called crude oil, petroleum is an extremely valuable resource that is used to make gasoline and other fuels, chemicals, and a wide variety of other products such as plastics.

Goals In this investigation, you will **measure** the amount of oil you can remove from a container by three different extraction methods. You will **observe** the effectiveness of each method and **graph** your results.

LAB WARMUP

Concepts Petroleum is found in underground deposits. The petroleum is usually held within porous rock layers, generally sandstone or limestone. Surrounding nonporous rock, such as shale or clay, tends to trap the oil and keep it from moving out of the porous rock. The petroleum can move, however, if an opening is made for it. It can rise to the surface when a well pipe is drilled down to the deposit, because natural gas or water that is trapped underground along with the petroleum exerts pressure on it. In such cases, extraction is quite easy. Usually, however, the petroleum must be extracted through a pumping process. Sometimes cold or hot water is forced down into the deposit to help dislodge the petroleum from the surrounding rock and move it up the well.

Review Section 15.3, Petroleum and Natural Gas, should be completed before beginning this investigation. You should also understand the following terms before you perform this investigation.

petroleum   extraction   pressure

Make a **prediction** about the outcome of this experiment and write it in the Lab Notebook.

MATERIALS (PER GROUP)

- rubber gloves
- plastic bottle with spray pump
- pebbles to fill the bottle halfway
- graduated cylinder
- 100-mL motor oil
- plastic tubing that fits pump nozzle
- 3 250-mL beakers
- cold tap water
- hot tap water

PROCEDURE

1. **CAUTION: Motor oil is poisonous and will also stain clothing. Wear rubber gloves, an apron, and goggles. Be careful not to spill any oil.** Half-fill the spray bottle with pebbles. Measure out 100 mL of motor oil and pour it into the bottle.

2. Screw the spray-pump top onto the bottle, making sure to work its tube down through the pebbles. Attach an external length of tubing to the nozzle and put the end of this tubing into the graduated cylinder. The setup is illustrated in Figure 20.1 below.

3. Press down repeatedly on the spray pump to remove as much of the oil as you can from the bottle and pump it to the cylinder. Use the graduated cylinder to measure the volume of oil removed. Record this information in the column of the Lab Notebook table labeled "Volume of oil actually removed." Also record your other observations, on the lines provided. Pour the oil back into a beaker.

4. Pour 80 mL of cold tap water into the spray bottle, which still contains some oil. Again try to pump out as much liquid as you can into another beaker. Let the resulting oil-water mixture separate out. Pour off the oil into the graduated cylinder, and record this volume in the Lab Notebook. Pour the oil back into the beaker. (For comparison purposes later, you can assume that the cold water would have been able to remove all the oil removed in step 3, plus the extra amount it actually removed.)

5. Repeat step 4, but this time add 80 mL of hot tap water instead of cold tap water. **CAUTION: Be careful not to burn yourself.** Again record the amount of oil you removed. Dispose all the oil in the container provided by your teacher for this purpose. (You can assume later that the hot water would have been able to remove all the oil removed in steps 3 and 4, plus the extra amount it actually removed.)

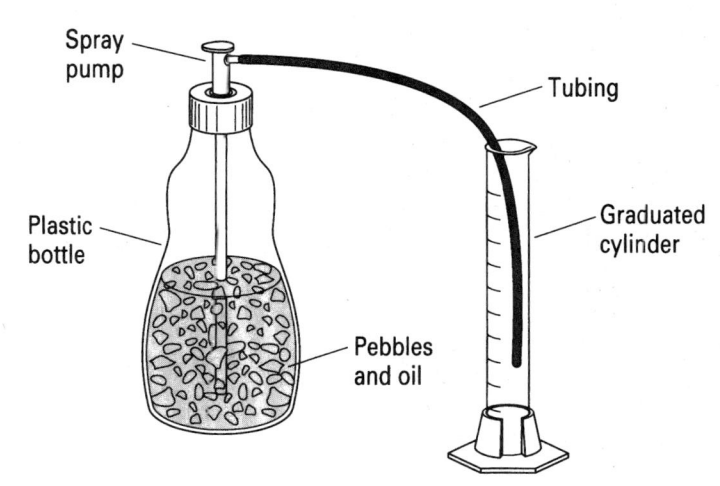

Figure 20.1 Setup of spray pump

Name: _____ Class: _____ Date: _____

LAB NOTEBOOK: INVESTIGATION 20

PREDICTION **A correct prediction is that the hot-water method of oil extraction will remove the most oil and the no-water method of oil extraction will remove the least oil.**

OBSERVATIONS

AMOUNTS OF OIL REMOVED BY VARIOUS METHODS

Pumping method	Volume of oil actually removed	Total volume of oil removable
No water		
Cold water	**Data and observations will vary, but the cold-water method will remove some oil left by the no-water method, and the hot-water method will remove some oil left by the cold-water method.**	
Hot water		

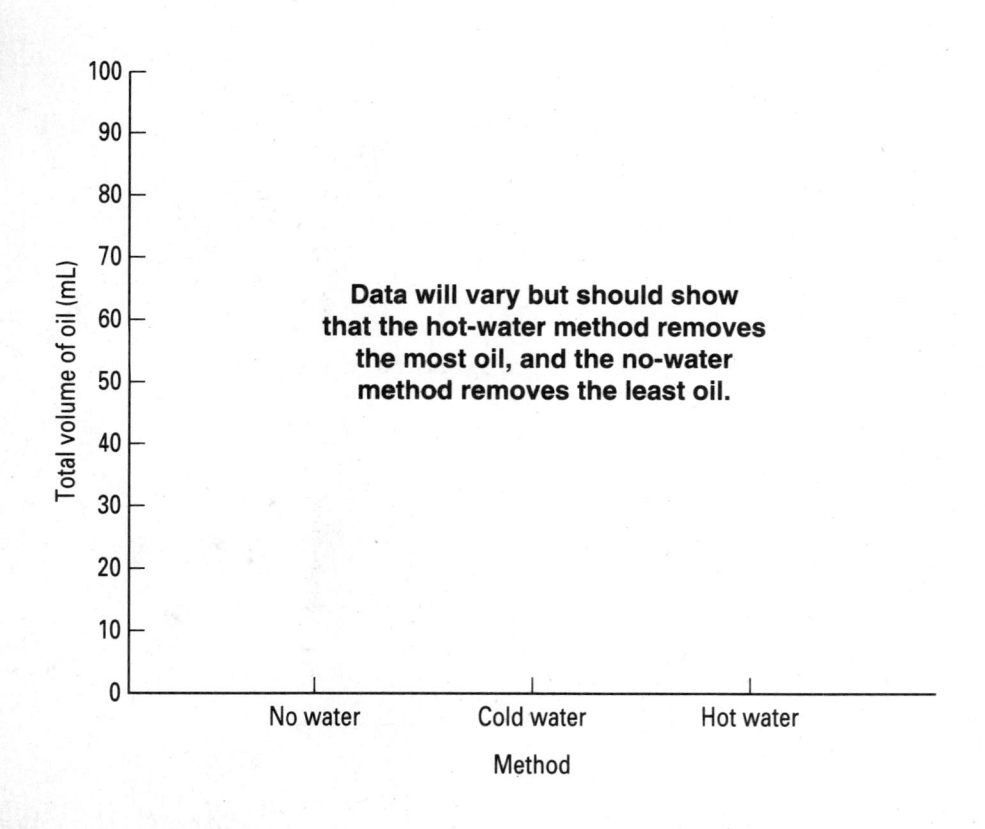

Data will vary but should show that the hot-water method removes the most oil, and the no-water method removes the least oil.

(Graph: Total volume of oil (mL) on y-axis from 0 to 100; Method on x-axis: No water, Cold water, Hot water)

DATA ANALYSIS

1. Add the volume of oil removed with the no-water method to the volume removed with the cold-water method. This total is the volume of oil that would have been removable by the cold-water method alone had the no-water method not already been used. Record this value in the column labeled "Total volume of oil removable," in the Observations table on the previous page.

2. Add the value you calculated in question 1 to the volume of oil removed with hot water. This total is the volume of oil that would have been removable by the hot-water method had the other two methods not been used. Record this value. Show your calculations below.

3. Using the grid in the Lab Notebook, make a bar graph comparing the total volume of oil that would have been removed using each oil-extraction method alone.

CONCLUSION

1. **Interpret data** Referring to your graph, compare the effectiveness of the three methods of oil extraction.
   The no-water method is the least effective in extracting oil. The hot-water method is the most effective.

2. **Infer** Explain why each of the methods was more or less effective.
   The no-water method was least effective because some of the oil clung to the pebbles and could not be removed by simple pumping. The cold water dislodged and displaced some of this remaining oil. The hot water made the oil more fluid and allowed it to be dislodged and removed more readily.

3. **Generalize** How are the methods and devices you used similar to those used in actual oil-drilling operations?
   The no-water method corresponded to the simple pumping of oil from a deposit. The pebbles corresponded to the porous, oil-containing rock, the spray-pump to the oil pump, the tubing to the well pipe. The cold- and hot-water methods corresponded, respectively, to the pumping of cold and of hot water into an oil deposit as part of an extraction method.

EXTENSION

Research Conduct library research to find out more about several oil extraction techniques and the benefits and problems associated with each. Write a report on your findings, and include diagrams or drawings illustrating the various devices and methods used.

LABORATORY INVESTIGATION

21 OBSERVING RADIATION

Problem: *How can cosmic radiation and nuclear radiation by observed?*

INTRODUCTION

Background All people are exposed to radiation from natural sources every day. For example, the Earth is constantly bombarded with cosmic rays caused by the sun and other sources. Naturally occurring radioactive elements, such as uranium, thorium, and radium, emit gamma radiation and help heat Earth's core and rocks. Radon, a decay product of radium, is a gaseous radioactive element present in the soil. About half of the radiation that we are naturally exposed to comes from radon.

Three different forms of radiation are emitted from the nuclei of radioactive atoms. *Alpha particles* are composed of two protons and two neutrons, and thus have a positive charge. These low-energy particles can penetrate air only, so they pose little danger to living organisms. *Beta particles* are high-speed electrons, which have a negative charge. They have more energy than alpha particles and can penetrate skin. However, beta particles can be stopped with a millimeter thickness of lead. *Gamma rays*, a form of electromagnetic radiation, are much more dangerous to living organisms than alpha or beta particles since they can penetrate farther.

Goals In this investigation, you will **construct** a cloud chamber and **observe** cosmic rays in the chamber. You will then add a source of alpha radiation to the cloud chamber and **observe** alpha particles in the chamber. You will then **compare** cosmic rays and alpha particles as seen in a cloud chamber.

LAB WARMUP

Concepts A *cloud chamber* is an apparatus used to observe radiation. The addition of alcohol and the cooling of the chamber create a dense alcohol-vapor layer in the chamber. As a cosmic ray or alpha particle passes through the cloud chamber, it ionizes molecules in its path. The ions attract alcohol molecules which then condense around them, forming droplets. The result is a white trail that displays the particle's path through the chamber. The quantity, direction, and appearance of these trails depend on the type of radiation observed.

The source of radiation you will use emits only alpha particles, which cannot penetrate the skin. Although the source if swallowed or inhaled can be dangerous, the radiation itself is harmless.

Review Section 16.1, Atoms and Radioactivity, should be completed before beginning this investigation. You should also understand the following terms before you perform this investigation.

alpha particles   beta particles   gamma rays   cosmic rays   cloud chamber

Make a **prediction** about the outcome of this experiment and write it in the Lab Notebook.

MATERIALS (PER GROUP)

- scissors
- black felt
- large glass jar with lid
- thick blotting paper
- fast-drying glue
- dropper
- isopropyl alcohol
- dry ice
- tray
- thin cloth rags
- flashlight
- radioactive needle inserted into a cork (or other source of alpha radiation)

PROCEDURE

1. Wear gloves, safety goggles, and your laboratory apron.
2. Construct a cloud chamber like the one shown in Figure 21. **CAUTION: Be careful when using scissors.** First cut the black felt in a circle slightly smaller than the bottom of the glass jar. Cut a circle of thick blotting paper to fit inside the lid of the jar.
3. Place the felt circle in the bottom of the jar.
4. Glue the circle of blotting paper to the inside of the lid.
5. **CAUTION: Be careful when handling alcohol, which is a flammable, toxic chemical.** Use the dropper to soak the blotting paper with alcohol. The blotting paper should be completely saturated but should not drip when held upside down.
6. Screw the jar onto the lid, and place the jar on the dry ice in the tray your teacher has provided for you. **CAUTION: Do not touch the dry ice.** Cover the dry ice with rags, wrapping them around the bottom of the jar, to avoid touching the ice and to keep it colder for a longer time.
7. In a somewhat darkened room, shine a flashlight into the jar from the side. Observe the jar for the next 5 to 10 minutes, recording your observations in the Lab Notebook.
8. Unscrew the lid of the jar. Add alcohol to the blotting paper if necessary. **CAUTION: Although alpha particles will not penetrate the skin, the source itself may be dangerous if swallowed or inhaled. If you touch the source, wash your hands.** Carefully place the radiation source on the bottom of the jar. If using a radioactive needle inserted into a cork, hold the cork while placing the source in the jar.
9. Screw the lid onto the jar.
10. In a somewhat darkened room, shine a flashlight into the jar from the side. Observe the jar for 10 to 15 minutes, recording your observations in the Lab Notebook. If the tracks become fuzzy, try rubbing the top of the jar in a circular motion with a dry rag. This should create static, attracting charged particles to the top of the jar and clearing the center of the jar.
11. Return the radiation source to your teacher when you have completed the experiment.

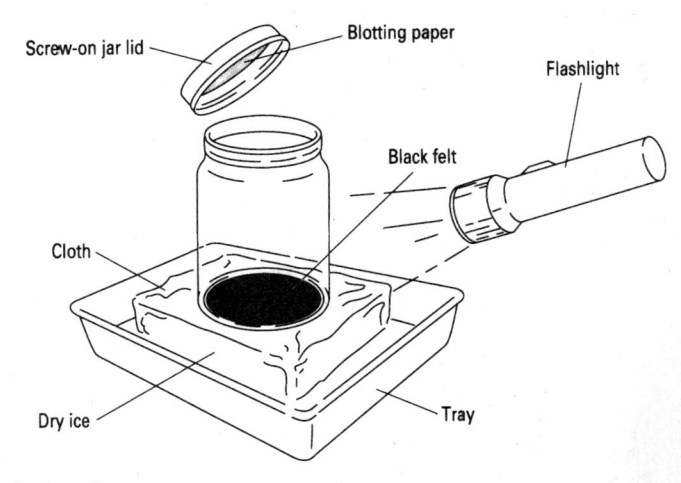

Figure 21 Setup of cloud chamber

Name: _____ Class: _____ Date: _____

LAB NOTEBOOK: INVESTIGATION 21

PREDICTION **A correct prediction is that cosmic rays can be observed as vapor trails in a cloud chamber. When a source of alpha radiation is added, the number of vapor trails increases.**

OBSERVATIONS

TRAILS PRODUCED BY COSMIC RAYS

Appearance _____

_____

Direction _____

Answers will vary. Trails should be

Length **random in direction and location and**

few in number.

Random or regular intervals _____

_____

Other observations _____

_____

TRAILS PRODUCED BY ALPHA PARTICLES

Appearance _____

_____

Direction _____

Answers will vary. Trails should be

Length **numerous and directed out from the**

radiation source.

Random or regular intervals _____

_____

Other observations _____

© Addison-Wesley Publishing Company, Inc. All Rights Reserved.

DATA ANALYSIS

1. Why was the cloud chamber placed on dry ice?
   The dry ice creates a temperature gradient inside the jar, making the alcohol vapor layer thick enough to see trails.

2. Was there more activity in the chamber before or after the addition of the alpha radiation source?
   There should be more activity in the chamber after the addition of the radiation source, assuming the source was emitting alpha particles as expected.

3. How do the trails in the chamber before and after the addition of the radiation source compare?
   The trails caused by cosmic rays should be more random than those caused by alpha particles. The trails caused by alpha particles should radiate from the source out into the chamber. All trails should be similar in appearance.

4. Explain the cause of what you observed in the cloud chamber during step 7 of the Procedure.
   Cosmic rays ionized particles in their paths; alcohol vapor condensed around these ions, forming vapor trails.

5. Explain the cause of what you observed in the cloud chamber during step 10 of the Procedure.
   Alpha particles produced ions in their path. Alcohol vapors condensed around the ions and formed a white trail.

CONCLUSION

1. **Infer** Are you constantly being exposed to cosmic radiation? Support your answer based on your observations in this investigation.
   Yes. The cloud chamber showed the presence of radiation without the addition of the alpha source.

2. **Predict** If sources of alpha and beta radiation were placed next to each other in a cloud chamber, with a positive plate on one side of the chamber and a negative plate on the other side, what might you observe? Explain your answer, drawing pictures on a separate sheet if necessary.
   The trails of the alpha particles (with a positive charge) would be directed toward the negative plate, while the trails of the beta particles (with a negative charge) would be directed toward the positive plate.

EXTENSION

Hypothesize Can the paths of alpha particles be altered? Design an experiment to test your answer to question 2 in the Conclusion section.

LABORATORY INVESTIGATION

22 THE EFFECTS OF RADIATION ON PLANTS

Problem: *How does radiation affect seed germination and plant growth?*

INTRODUCTION

Background Excessive exposure to nuclear radiation can be harmful to all living things. Ionizing (high-energy) radiation, such as gamma rays, can alter the genetic material of cells, causing *mutations*. If the irradiated organism is in an early stage of development, the mutated cell and all cells arising by cellular divisions from that cell will carry the mutation. If the mutation is harmful, the organism may be malformed or may fail to develop. Because of the potential dangers involved, many people question the safety of using nuclear energy. If the radiation is not well contained, if an explosion occurs in a nuclear reactor, or if nuclear wastes are not properly disposed of, all living things in the area could be exposed to radiation.

Goals In this investigation, you will plant barley seeds that have been exposed to different levels of radiation. You will **observe** the germination of the seeds and the growth of the seedlings, and **infer** how radiation affects plant growth and development.

LAB WARMUP

Concepts Scientists use a unit called a *rad* to describe radiation dosage. A rad, or "radiation absorbed dose," is the amount of radiation absorbed by something. Specifically, it is the absorption of 100 ergs of energy per gram of tissue (an erg is a small unit of energy). A *rem* (rad equivalent man) is the amount of radiation absorbed by humans. One rem is equal to a dosage of one rad of X rays to a person. The seeds you will use in this experiment have been exposed to a tremendous amount of gamma radiation (20,000 to 50,000 rads). A dose of 1000 rads would kill 100 percent of humans within a month of exposure. Note that the seeds you will use are not radioactive; they do not emit ionizing radiation, so they are safe to handle.

Review Section 16.3, Radioactive Waste, should be completed before beginning this investigation. You should also understand the following terms before you perform this investigation.

ionizing radiation   gamma rays   X-rays   radiation dose   rad

Make a **prediction** about the outcome of this experiment and write it in the Lab Notebook.

MATERIALS (PER GROUP)

- 5 pairs of rubber gloves
- planting flat
- masking tape
- marker
- potting soil, moistened
- hammer
- 8 nails or tacks
- string
- water
- scissors
- 5 packets of barley seeds, including nonirradiated seeds and seeds irradiated with four different dosages (20,000, 30,000, 40,000 and 50,000 rads)
- metric ruler
- large transparent plastic bag with tie
- graph paper
- colored pencils

PROCEDURE

1. Put on gloves and your laboratory apron.
2. Obtain a planting flat and, using masking tape, label it with your group's name. Fill your planting flat with moistened soil.
3. Divide the flat into five rows, and label the end of each row with one amount of radiation dosage. **CAUTION: Be careful when using a hammer and nails. Wear safety goggles when doing this.** To mark the boundaries of the rows, hammer nails or tacks into the sides of the flat, and tie pieces of string between them.
4. Count the number of barley seeds in the first packet of seeds. Record this number in the first table of the Lab Notebook under "Number of seeds planted" next to the appropriate radiation dosage.
5. Evenly scatter the seeds in the appropriate row in the flat, making sure the seeds do not fall outside the boundaries of the row. Push the seeds down into the soil to about 5 mm in depth, and then cover them with a thin layer of soil.
6. Repeat steps 4 and 5 for the remaining packets of seeds.
7. Place the flat in a transparent plastic bag and tie the end of the bag. Set the flat at room temperature in a well-lighted area, but not in direct sunlight.
8. When the first seedlings appear, remove the flat from the plastic bag. For each radiation dosage, record the date when the first seedling appears. Check the soil every other day, making sure it remains moist but not saturated.
9. One week after the planting date, count the number of seeds that have germinated for each level of radiation exposure. Record these data in the appropriate spaces in the Lab Notebook. For each dosage, calculate the percentage of seeds that have germinated:

$$\text{percent gemination} = \frac{\text{number of seedlings}}{\text{total number of seeds planted}} \times 100$$

Record the percentages in the appropriate spaces in the Lab Notebook.
10. Measure the height of each seedling in each row, calculating an average seedling height for each level of radiation exposure. Record the averages in the appropriate spaces in the Lab Notebook.
11. Record your observations about the seedlings' leaves, stems, colors, and overall appearance in the second table of the Lab Notebook.
12. Repeat steps 9, 10, and 11 two weeks and three weeks after the planting date.
13. Using a different colored pencil for each radiation dosage, graph your gemination data on a separate sheet of graph paper. Plot time on the x-axis (in numbers of weeks) and percent germination on the y-axis. Label the dosage represented by each color.

Name: _____ Class: _____ Date: _____

LAB NOTEBOOK: INVESTIGATION 22

PREDICTION _A correct prediction is that exposure of seeds to radiation inhibits their germination and causes the seedlings to be stunted and malformed._

OBSERVATIONS

Radiation absorbed dose (rad)	Number of seeds planted	Germination date of first seed	Number of seeds germinated Week 1	2	3	Percent germination Week 1	2	3	Average seedling height Week 1	2	3
No radiation											
20,000											
30,000	Data will vary. Increasing levels of radiation should show decreasing percent germination and seedling height.										
40,000											
50,000											

SEEDLING OBSERVATIONS

Radiation absorbed dose (rad)	Week 1	Week 2	Week 3
No radiation			
20,000			
30,000	Data will vary. Increasing levels of radiation should show more evidence of malformation.		
40,000			
50,000			

© Addison-Wesley Publishing Company, Inc. All Rights Reserved.

DATA ANALYSIS

1. What was the purpose of the nonirradiated seeds used in this experiment?
   They serve as the control of the experiment, to which the irradiated seeds can be compared.

2. Which group of seeds showed the earliest germination? The latest?
   Answers may vary, but the nonirradiated seeds will probably germinate the earliest; the seeds exposed to 50,000 rads will germinate the latest.

3. How do increasing levels of radiation exposure affect the percent germination of the barley seeds?
   Fewer seeds germinate with increasing radiation exposure.

4. How do the increasing levels of radiation exposure affect the growth of the barley seedlings?
   The seedlings are more stunted with increasing radiation exposure.

5. How do highly irradiated seedlings compare to nonirradiated seedlings in appearance?
   Answers will vary, but the irradiated seedlings may have more fragile stems, malformed leaves, etc.

CONCLUSION

1. **Infer** Ionizing radiation affects plants at the molecular level. It alters the genetic material that directs the formation of proteins which control growth and development. Explain why radiation exposure of seeds produces more dramatic results than exposure of mature plants.
   Radiation exposure of seeds may mutate cells in such a way that they are unable to divide, or if they do so, produce defective proteins. When a mutated cell does divide, each new cell makes the same defective protein. Thus, the seed may not germinate at all or may produce a defective seedling.

2. **Infer** Why do hospitals post signs asking "Are You Pregnant?" in areas where X ray are used?
   X rays are more harmful to developing fetuses than to adults. Exposure may result in defects in the unborn child.

EXTENSION

Hypothesize How does ionizing radiation affect plants that produce fruit? Repeat the experiment, using irradiated tomato seeds provided by your teacher. Remember to use a nonirradiated control. Grow the plants until they bear fruit. Compare the effects of varied levels of radiation exposure on the structure, color, and size of leaves, flowers, and fruit. Plant the seeds in a school garden or use a separate window box for each seed type, since tomato plants require more room to grow to maturity.

LABORATORY INVESTIGATION 23

SOLAR ENERGY CONCENTRATOR

Problem: *How can solar energy by concentrated? What position maximizes the efficiency of a solar concentrator?*

INTRODUCTION

Background   There are many benefits in using solar energy to help meet human energy needs. The use of solar energy does not deplete Earth's limited resources. Harnessing solar energy does not pollute the environment, and sunlight costs nothing. These advantages make solar energy an excellent alternative to other energy sources such as fossil fuels.

Solar collectors can be used in place of furnaces to heat homes and other buildings. A solar collector is usually a flat box with a black metal base and a glass cover. Fluid-filled tubes are mounted on the base. The fluid gives off heat as it circulates through the building. Then it returns to the solar collectors to be reheated. Unfortunately, very little of the sun's total energy is actually harnessed with a solar collector. More solar energy, however, can be collected with another device called a solar energy concentrator. In the concentrator, reflective material is used to direct the sun's rays onto a central point. The collector is in a shape called a *parabola*. A parabola is a curve that reflects all parallel rays striking it toward a point called the *focal point*.

Goals   In this investigation, you will **build** a solar energy concentrator. You will then **test** its ability to heat water and **record** the temperature rise in the water over time. You will **compare** the effectiveness of the concentrator in two different positions.

LAB WARMUP

Concepts   The type of concentrator you will build for this experiment is a parabolic solar energy concentrator. The reason for using a parabola is to direct the sun's rays toward a focal point.

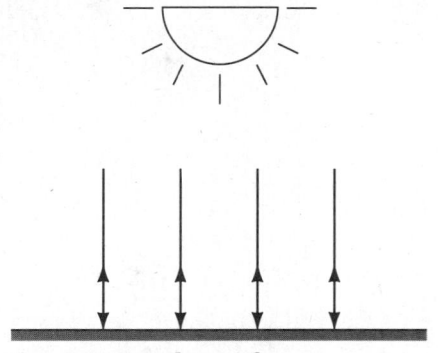

Figure 23.1   Flat surface

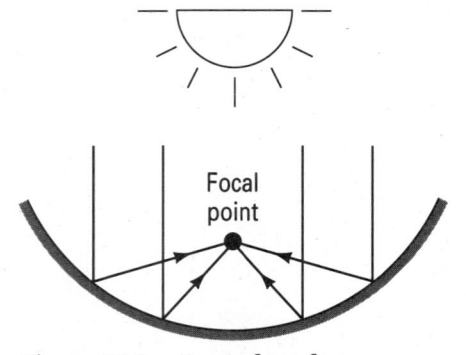

Figure 23.2   Curved surface

Notice the curved surface in Figure 23.2. Observe the difference a curved surface makes in directing the sun's rays. Figure 23.2 shows a cross section of the parabolic concentrator you will build. Rather than directing the rays toward a single point (as in the drawing), the rays will be directed along a line that extends the length of the concentrator.

Review   Section 17.1, Solar Energy, should be completed before beginning this investigation. You should also understand the following terms before you perform this investigation.

solar energy collector   solar energy concentrator   parabola

Make a **prediction** about the outcome of this experiment and write it in the Lab Notebook.

© Addison-Wesley Publishing Company, Inc. All Rights Reserved.

97

MATERIALS (PER GROUP)

- 2 aluminum sheets
- 4 precut wooden ends
- tacks
- 2 hammers
- 2 pieces of glass tubing
- 2 rubber stoppers
- water
- 2 thermometers
- 2 rubber rings to fit around thermometers
- colored pencils

PROCEDURE

1. Your group will be divided into two smaller groups. Each group will be responsible for assembling one parabolic solar energy concentrator and for recording temperature readings from the concentrator. **CAUTION: Be careful and wear safety goggles when using a hammer and tacks.** Attach a piece of shiny aluminum to two wooden ends with tacks and a hammer, as shown in Figure 23.3.

2. **CAUTION: Be careful when handling glassware. Mercury is poisonous. Do not touch it if your thermometer breaks. Notify your teacher immediately.** Fit a rubber ring around the thermometer, about one third of the way from the tip. Insert the thermometer into the tubing until the rubber ring fits into it. The seal should be snug but not airtight.

3. Fill the glass tubing with water.

4. Seal the other end of the glass tubing with a stopper.

5. Record the temperature of the water in the Lab Notebook.

6. Carefully insert the tube/thermometer apparatus through the holes of the wooden ends of the collector.

7. One group should then place their solar energy concentrator flat on the ground in the sun. The other group should place theirs in the sun, but tilting the trough such that it points directly toward the sun. Prop the concentrator in place with a book or other object with the thermometer end raised for the duration of the experiment.

8. Record the temperature readings of each thermometer every 5 minutes for 60 minutes, if time allows. **CAUTION: Remove the apparatus from the sun if the temperature reaches 85° C. Let the apparatus cool before disassembling it.**

9. Plot these data on the graph in your Lab Notebook. Use a different color for each set of data. Connect the same-colored points, and label each line on the graph either "indirect sunlight" or "direct sunlight."

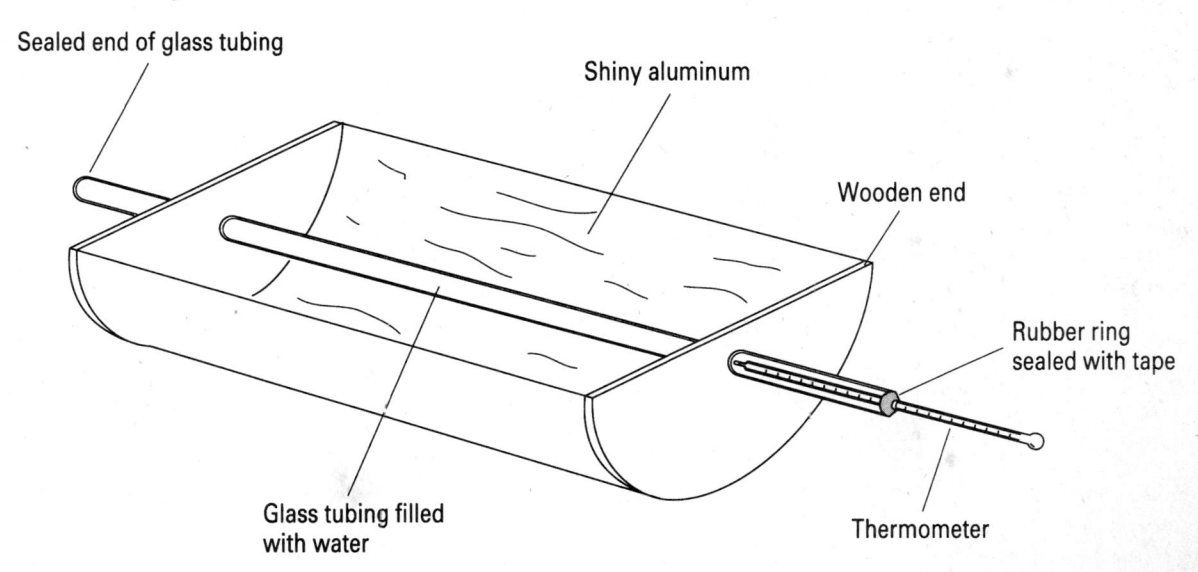

Figure 23.3 Setup of solar energy concentrator

Name: _____ Class: _____ Date: _____

LAB NOTEBOOK: INVESTIGATION 23

PREDICTION <u>A correct prediction is that a solar concentrator pointing directly toward the</u>
<u>sun will heat water more quickly than one not directly pointing toward the sun.</u>

OBSERVATIONS

TEMPERATURE READINGS FOR SOLAR ENERGY CONCENTRATORS

Time (min)

	0	5	10	15	20	25	30	35	40	45	50	55	60
Indirect sun													
Direct sun													

Data will vary but should show that the water is heated more quickly in direct sunlight.

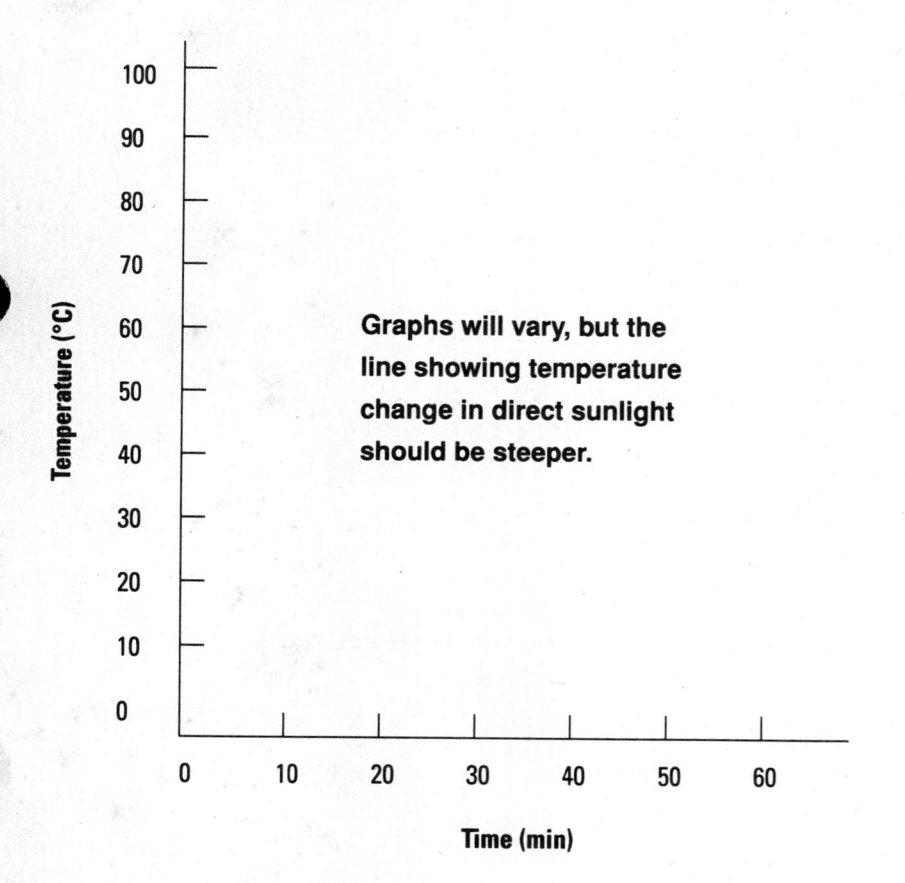

Graphs will vary, but the line showing temperature change in direct sunlight should be steeper.

(Graph: Temperature (°C) vs. Time (min), 0–100 °C by 10, 0–60 min by 10)

© Addison-Wesley Publishing Company, Inc. All Rights Reserved.

DATA ANALYSIS

1. Why is the shape of the solar energy concentrator parabolic?
   The parabolic shape allows rays to converge on a certain area.

2. What happened to the temperature of the water in each concentrator over time?
   The temperature of the water increased.

3. Which concentrator had a higher final temperature reading?
   The concentrator tilted directly toward the sun will show a higher temperature reading.

4. From the graph, does the water ever reach a maximum temperature, which then levels off? If so, what is the maximum temperature?
   Answers will vary.

CONCLUSION

1. **Infer** What is the most efficient position for a solar energy concentrator?
   The trough of the solar energy concentrator should be directed toward the sun.

2. **Infer** If a solar energy concentrator was placed on the roof of a house, how could its efficiency be maximized throughout the day?
   The concentrator would have to move throughout the day to adjust to the movement of the sun from east to west.

3. **Predict** What other designs might make the solar energy concentrator more efficient?
   Accept all logical responses. For example, students might suggest adding mirrors or making the concentrator dome-shaped.

EXTENSION

Predict How will a cloudy day affect the solar concentrator? Repeat the experiment on a cloudy day to test your prediction. Compare your results to those obtained in this lab. Is solar energy reaching the concentrator and heating the water?

LABORATORY INVESTIGATION

24 ELECTRICITY FROM SOLAR CELLS

Problem: *How can the sun be used as an energy source?*

INTRODUCTION

Background A *photovoltaic cell*, or solar cell, converts solar energy to electricity. When a photon of light enters the cell, it knocks an electron free from one of the atoms in the cell. A neighboring electron takes its place. The space left by the neighboring electron is then filled by another, and so on. This movement of electrons causes an electrical potential difference, or voltage, to build up in the cell. If the cell is connected to a circuit, it produces a current through the circuit.

Solar cells consist of layers of materials called *semiconductors*, which are capable of transmitting electric current, but not as efficiently as metals such as copper. The most commonly used semiconductor material is silicon. Half of the layers in the cell are coated with a substance such as arsenic or phosphorous having a large supply of loosely bound electrons. Between these layers are layers coated with a substance such as boron, which has an excess of positive charge to which the freed electrons can move.

Solar cells are not yet very efficient. About one-sixth of the solar energy that hits the cell is converted to electricity. Scientists are attempting to develop more efficient solar cells, and some experimental cells have an efficiency rate of 30% or more. The low efficiency of solar cells prevents their use in large-scale power plants, since several square miles of land covered with costly solar cells would be needed to produce the same amount of power that an average power plant would.

Goals In this investigation, you will **measure** the voltage and current produced by a solar cell, and **evaluate** solar energy as an alternative to nonrenewable energy sources.

LAB WARMUP

Concepts When a power supply such as a solar cell is connected to a closed circuit, measurements of the voltage and the current can be made. *Voltage* is the electrical potential difference across the solar cell, caused by freed electrons moving through the cell. A *voltmeter* measures this potential in volts. An *ammeter* measures the current, or the rate of movement of electrons passing through the circuit. Current is expressed in units of *amperes*. The resistor you will use in your test circuit conducts some electricity, but also dissipates some as heat, similar to the action of heater elements and light bulb filaments. The unit of resistance is the ohm, which is volts per ampere.

The efficiency of a solar cell can be estimated by making a number of calculations. The power output of the cell is a function of the voltage and current produced by the cell:

power output (watts) = voltage (volts) × current (amps)

The power density of the cell, or how powerful the cell is for its size, is calculated:

$$\text{power density (watts/cm}^2\text{)} = \frac{\text{power output (watts)}}{\text{area (cm}^2\text{)}}$$

The efficiency of the cell is then expressed as a percentage ratio of the power density of the cell and the amount of solar energy falling on the cell:

$$\text{efficiency (\%)} = \frac{\text{power density}}{\text{solar input}} \times 100\%$$

Review Section 17.1, Solar Energy, should be completed before beginning this investigation. You should also understand the following terms before you perform this investigation.

photovoltaic cell   potential difference   current   ammeter   voltmeter   volt   ampere

Make a **prediction** about the outcome of this experiment and write it in the Lab Notebook.

MATERIALS (PER GROUP)

- 2 lead wires with alligator clips
- voltmeter (high resistance), 0-1V
- milliammeter (0-100mA)
- 20- to 22- ohm resistor
- solar cell
- metric ruler

PROCEDURE

1. Build a test circuit as shown in Figure 24.1, using lead wires, a voltmeter, a milliammeter, and a resistor.
2. Connect a solar cell to the circuit shown in Figure 24. Place the circuit in direct sunlight.
3. Observe the current flow (I) in milliamperes and the electric potential (V) in volts. Record these data in the Lab Notebook.
4. Calculate the power output of the solar cell by multiplying current in amps by electric potential in volts. Record the power output in the Lab Notebook.
5. Measure and record the length and width of the solar cell (the inside black part, not the plastic casing). Do not open the casing.
6. Calculate the power density of the solar cell by dividing the power output of the cell by the area of the cell. The power density of a solar cell will enable you to determine how many power cells are needed to meet a certain energy requirement.
7. Calculate the efficiency of the solar cell as an energy converter. Approximate the solar input at your elevation (solar input at sea level = 0.08 watts/cm^2; solar input at 1500 m = 0.10 watts/cm^2). The efficiency is the ratio of the power density of the solar cell to the solar input. Express this value as a percentage.

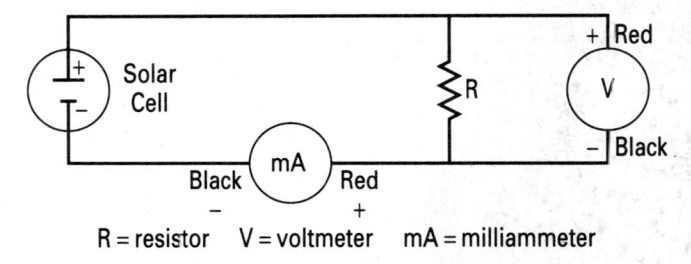

Figure 24.1 Construction of the test circuit

Name: _____ Class: _____ Date: _____

LAB NOTEBOOK: INVESTIGATION 24

PREDICTION <u>A correct prediction is that the wind and the sun can be used to produce electricity.</u>

OBSERVATIONS

SOLAR CELL DATA

Current = _____ mA = _____ amps.   **Data will vary.**

(1000 mA = 1 ampere)

Electric potential = _____ volts

Power output = _____ amps × _____ volts = _____ watts

Length of cell = _____ cm

Width of cell = _____ cm

Area of cell = length × width = _____ cm × _____ cm = _____ cm²

$$\text{power density (watts/cm}^2\text{)} = \frac{\text{power output (watts)}}{\text{area (cm}^2\text{)}} = \_\_\_\_\_ \text{ watts / cm}^2$$

Efficiency of solar cell as energy converter:

$$\text{efficiency (\%)} = \frac{\text{power density}}{\text{solar input}} \times 100\% = \_\_\_\_\_ \%$$

© Addison-Wesley Publishing Company, Inc. All Rights Reserved.

DATA ANALYSIS

1. How many solar cells of this size would you need to light a 75-watt bulb at this time of day?

   Answers will vary, depending on the power output. Divide 75 by the power output of one cell to calculate the number of cells needed.

2. Electrical energy for consumer purposes is measured in kilowatt-hours, which is the power in kilowatts multiplied by time in hours (1 kw = 1000 watts). Assume your home requires 15 kwh per day for light, heat, and electrical appliances. Your home's energy needs are met by a solar cell system. There is an average of 6 hours of effective collecting time per day and energy storage is available. What is the required power output of the solar cell system for your home?

   The solar cell system needs to produce 2.5 kilowatts to meet household needs.

3. How many solar cells of the type you tested would the home need on its roof to meet these energy requirements?

   Answers will vary. Divide 2500 watts by the power output of one square cm of cell to find the area in cm².

CONCLUSION

1. If the cost of installing a solar cell system is $5000 per kilowatt, how much would it cost to install a system to meet your household needs?

   It would cost 2.5 × $5000, or $12,500.

2. If the cost of electricity in your area is 10¢ per kWh, how much would it cost to meet the needs of the house for a 30-year period? How does this compare with the cost of installing the solar cell system?

   It would cost 10¢ × 15 kWh/day × 365 days × 30 years = $16,425, which is more expensive than the cost of installing a solar cell system.

3. **Generalize** Discuss the benefits and drawbacks of solar energy. Could solar energy be used successfully in all areas of the United States?

   Answers will vary. Benefits include the availability of a free solar energy and its renewability. Drawbacks include the cost of initial installation and a need for storage for times when the energy source is not available. Solar energy would not be practical in the far north or in areas with frequent and heavy cloud cover.

EXTENSION

Research How would a higher efficiency rating of photovoltaic cells affect the possibility of large-scale solar power plants? If a solar cell with a 40% efficiency rating was developed, research and discuss the effects it would have on the area of solar cells needed and the cost of building such a facility. Compare this to the possibility of alternative sources of power, such as wind power and nuclear power.

LABORATORY INVESTIGATION

25 SOIL EXPLORATION

Problem: *How do different soils in your area compare in terms of their composition and water drainage?*

INTRODUCTION

Background Soil is made up of weathered, or broken-down, particles of bedrock and living and decaying organisms. Soils differ in their composition and the size of their particles. Soils also differ in the rate at which water drains through them, and in how well they hold water. A soil's ability to hold water is called its *water retention*.

The composition and size of soil particles determine the type of soil. Soil type, in turn, determines the numbers and kinds of organisms that live in the soil. Thus, different kinds of plants and animals in an ecosystem are supported by different soil types. For example, desert plants survive best in fast-draining soil. Plants that require larger amounts of water survive best in soil that holds more water and drains more slowly.

Goals In this investigation, you will **compare** different kinds of soils. You will **observe** a soil sample taken from an area near your school, **measure** the rate at which water drains through the soil, and **classify** the soil. Then you will **compare** your soil sample with those of your classmates.

LAB WARMUP

Concepts If particles in a soil are less than 0.002 mm in diameter, they cannot be distinguished without a microscope. If they range from 0.002 mm to 0.05 mm, they can be seen through a magnifying glass, but not without one. If the soil particles are larger than 0.05 mm, they can be easily distinguished by eye. Soil may contain one size of particle or a mixture of all three sizes. Based on these and other properties, soils can be classified as *sand*, *silt*, *clay*, or *loam*, as you will learn in this investigation.

Review Section 18.3, Soil and Its Formation, should be completed before beginning this investigation. You should also understand the following terms before you perform this investigation.

loam   silt   clay   water retention

Make a **prediction** about the outcome of this experiment and write it in the Lab Notebook.

MATERIALS (PER GROUP)

- rubber gloves
- labeled samples of sand, silt, clay, and loam soils
- magnifying glass
- sample of soil collected near your school
- styrofoam cup
- paper towels
- cotton
- metric ruler
- graduated cylinder
- beaker with graduations
- clock or watch

© Addison-Wesley Publishing Company, Inc. All Rights Reserved.

PROCEDURE

1. **CAUTION: Wear gloves and an apron when handling soil.** Examine the samples of sand, silt, clay, and loam soils provided by your teacher. Note whether the particles in each soil are readily visible to the eye, readily visible only through a magnifying glass, or not visible even through a magnifying glass. Record your observations in your Lab Notebook. Moisten each soil slightly, and observe whether it feels grainy or sticky and whether it can be molded easily with your fingers. Record this information as well.

2. As directed by your teacher, collect soil from a specific site near your school. Other groups of students will collect their samples from other nearby sites. Collect enough soil to half-fill a styrofoam cup.

3. Examine the soil you have collected by spreading it out on paper toweling. Note whether particles are visible by eye, through a magnifying glass, or not at all. Also note whether organisms are present in it, and attempt to identify them. Record your observations in the first column of the chart in the Lab Notebook.

4. Moisten a small amount of the soil, reserving the rest for the next step. Note whether the moistened soil feels grainy or sticky, and how readily it molds. Record this information in the Lab Notebook.

5. Use a pencil or pen to make a 1-cm-diameter hole in the center of the bottom of the styrofoam cup. Place a small piece of cotton over the hole. Put into the cup some of the reserved dry soil you collected. Gently pack this soil down, forming a layer that is 5 cm thick. Remove any extra soil. Make sure there are no spaces between the edge of the soil and the cup.

6. Pour 100 mL of tap water into a graduated cylinder. Hold the styrofoam cup containing the soil over a beaker that has volume calibrations. Do not squeeze the cup.

7. Gently pour the water from the graduated cylinder into the styrofoam cup. Record the amount of time it takes for the first drop of water to come through the hole in the bottom. Record the volume of water that has collected in the beaker after 2 minutes and then again after 5 minutes.

8. Exchange data with groups of students who collected soil from different sites. Copy their data into the Lab Notebook.

Name: _____ Class: _____ Date: _____

LAB NOTEBOOK: INVESTIGATION 25

PREDICTION __A correct prediction is that clay soils have the smallest particles, the slowest water drainage, and the greatest water retention. Sandy soils have the largest particles, the fastest water drainage, and the least water retention.__

OBSERVATIONS

CHARACTERISTICS OF FOUR KINDS OF SOILS

Soil	Visibility of particles	Feel when moistened	Ease of molding
Sand	easily, by eye	grainy	difficult
Silt	by magnifying glass	somewhat grainy	moderate
Clay	not visible at all	sticky	easy
Loam	combination of above	combination of above	moderate

CHARACTERISTICS OF COLLECTED SOIL SAMPLES

	1	2	3	4	5	6	7	8
Visibility of particles								
Organisms present		**Data will vary, depending on soil type collected.**						
Feel when moistened								
Ease of molding								
Time for first drop								
Water volume after 2 min								
Water volume after 5 min								

© Addison-Wesley Publishing Company, Inc. All Rights Reserved.

DATA ANALYSIS

1. Based on their characteristics, classify the soil samples collected by your group and the other groups of students whose data you examined.

   Sample 1: **Answers will vary, but classi-** Sample 5: **Answers will vary, but classi-**
   Sample 2: **fication depends on the** Sample 6: **fication depends on the**
   Sample 3: **soil's particle size, feel, and** Sample 7: **soil's particle size, feel, and**
   Sample 4: **ease of molding.** Sample 8: **ease of molding.**

2. Compare the rate at which the different kinds of soil drained water. Which kind of soil drained fastest? Slowest?
   Sandy soil drains fastest, clay soil slowest.

3. How much water did your collected soil sample retain after 5 minutes? (Subtract the amount of water that drained out of the soil from 100 mL, the amount added to the soil.) Carry out the same calculation for the samples collected by other students. How did the total amounts of water retained by the different kinds of soils compare?
   Answers will vary. Sandy soil retains the least water, clay soil the most.

CONCLUSION

1. **Infer** Based on the characteristics of the different kinds of soils and what you know about the needs of most plants, in which kind of soil would desert plants grow best? Which kind would be best for most other types of plants? Explain your answer.
   Desert plants, which require little water, would do best in fast-draining sandy soil. Most plants require moderate amounts of water and reasonably good drainage, and thus would do best in silt or loam soil.

2. **Infer** How might the presence or absence of organic matter help to determine the number of organisms in soil? How might the characteristics of soil affect the amount of organic matter?
   Soils, such as loam, that support abundant plant life would have more organic matter from decaying plants. This material helps the soil retain water.

EXTENSION

Classify Expand your investigation of local soils by collecting samples from different places in and around your community. Obtain permission from property owners, and have an adult accompany you during the collections. Combine your data with those of other students and prepare a map of your area, coded to show the kinds of soils present.

LABORATORY INVESTIGATION

26 LANDFILL BIODEGRADATION

Problem: *How do the rates of decomposition compare for different substances placed in a model landfill?*

INTRODUCTION

Background   Solid wastes consist of garbage and by-products from agriculture and other human activities. In the United States, open dump sites were used for disposal of such wastes until the mid-20th century. Such sites are still used in most of the rest of the world. Open dumping, however, contributes to water and air pollution. It also causes the release of foul-smelling gases, as different organic and inorganic materials break down. Open sites are also unsightly as well as unhealthful for those who live near them. Materials dumped in open sites are not usually compacted. A great deal of valuable land space is therefore taken up.

Other forms of waste disposal, such as landfill, have now replaced most open dumping in the U.S., but they present a new set of environmental challenges. A landfill is a large pit usually lined with plastic. Pipes run under the pit to collect liquid that runs from the trash and penetrates the plastic. Trash is dumped and compacted, and then covered with a thin layer of soil. After the pit has been filled, the landfill is covered with several feet of soil. More pipes are run down into the trash to allow gases to vent.

Goals   In this investigation, you will **model** a landfill and **observe** changes in the materials that you placed in the landfill. You will then **infer** the relative decomposition rates of those materials and the problems they may present as part of the solid waste stream.

LAB WARMUP

Concepts   Currently, most garbage is dumped in landfills. At the landfill site, wastes are compacted, layered with soil, and buried. The degree to which materials are able to be broken down is a measure of how suitable they are for disposal in landfills. Biodegradable materials, those that decompose easily and enrich the soil, are the most suitable for inclusion in landfills. Nonbiodegradable materials last much longer in the environment and should be disposed of some other way, such as recycling. There is evidence that even biodegradable materials do not break down in landfills because air cannot reach the buried trash. It is therefore best to recycle or compost as much waste as possible.

Review   Section 19.1, Solid Wastes, should be completed before beginning this investigation. You should also understand the following terms before you perform this investigation.

landfill   decompose   biodegradable   nonbiodegradable

Make a prediction about the outcome of this experiment and write it in the Lab Notebook.

MATERIALS (PER GROUP)

- rubber gloves
- 1-L beaker
- garden soil
- 1 small piece of each of the following materials: glass, paper, aluminum, orange peel, apple, bread, wood, and plastic
- centimeter ruler
- water
- aluminum foil
- glass rod

PROCEDURE

1. **CAUTION: Wear gloves while handling the soil and objects.** Fill a large beaker three-fourths full with garden soil.
2. Obtain the eight objects made of different materials. Measure the length and width of each to the nearest millimeter, using a metric ruler. Record this information in the Lab Notebook. Also note each object's color and texture (smooth, rough, porous, hard, soft, dry, mushy, grainy, etc.)
3. Bury the eight objects at different points next to the inner surface of the glass, so that you can see them from the outside. Make sure they are all the same depth of soil, as shown in the figure below.
4. Thoroughly and evenly moisten the soil, but do not saturate it. Cover the beaker with foil.
5. Observe the objects in the beaker at the end of 1 week. In the Lab Notebook, describe any changes in their color and texture. You may probe them gently with a glass rod to make it easier to determine texture. Use the ruler to measure their size, and record this information.
6. Repeat step 5 at 1-week intervals for the following 3 weeks. Add water to the beaker to keep the soil moist, but do not saturate the soil.

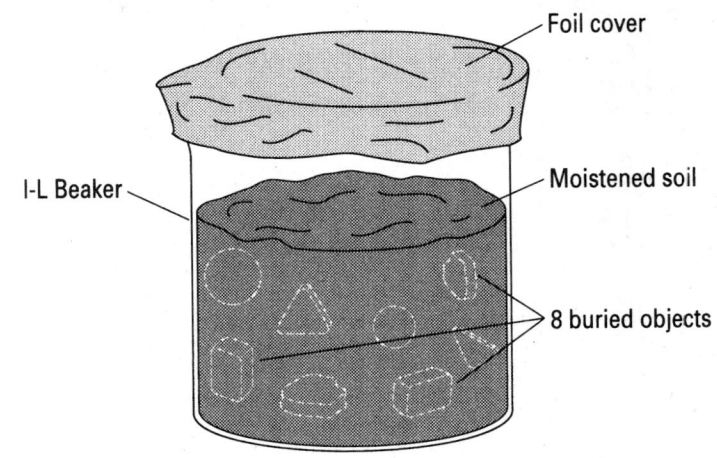

Figure 26.1 Model landfill

Name: _____ Class: _____ Date: _____

LAB NOTEBOOK: INVESTIGATION 26

PREDICTION __A correct prediction is that foodstuffs will decompose most rapidly; metal, glass, and plastics will decompose least rapidly.__

OBSERVATIONS

SIZE AND APPEARANCE OF MODEL LANDFILL

OBJECTS

Object	At start			After 1 week			After 2 weeks			After 3 weeks			After 4 weeks		
	Size	Color	Texture	Size	Color	Texture	Size	Color	Texture	Size	Color	Texture	Size	Color	Texture
Glass															
Paper															
Aluminum															
Orange peel															
Apple															
Bread															
Wood															
Plastic															

© Addison-Wesley Publishing Company, Inc. All Rights Reserved.

DATA ANALYSIS

1. Which materials changed considerably during the observation period? What did these materials have in common?
   The orange peel, apple, and bread changed considerably. They were all foodstuffs produced from living things.

2. Which materials changed slightly during the observation period? What did these materials have in common?
   The paper and wood changed slightly. They were plant materials, but were not foodstuffs.

3. Which materials did not seem to change at all during the observation period? What did these materials have in common?
   The glass, aluminum, and plastic remained unchanged. They were not derived from living things.

4. Why do you think it was important to keep the soil moist during this investigation?
   Moisture aids in the growth of bacteria, which are responsible for decomposition.

CONCLUSION

1. **Infer** Among the materials you studied, which would be most suitable for disposal in a landfill? Explain your answer, referring to the concept of biodegradability.
   The foodstuffs would be most suitable in that they are biodegradable—that is, they break down readily and produce soil nutrients. Food wastes should be composted, however, not placed in landfills.

2. **Infer** Among the materials you studied, which would be least suitable for disposal in a landfill? Explain your answer. Are the materials biodegradable? What problems would be created by disposal of such materials?
   The glass, aluminum, and plastic would be least suitable in that they are not biodegradable—that is, they do not break down readily or produce soil nutrients. Such materials continue to take up large amounts of space in landfills for very long periods of time and in some cases produce harmful substances. They should be recycled.

EXTENSION

Integrate Most landfills leak liquid or other pollutants eventually. Design a landfill that would be safe from leakage for several hundred years. Your design must let in air for decomposition and must not cost a lot of money. Are there better options than landfilling?

LABORATORY INVESTIGATION 27

TESTING WATER QUALITY

Problem: *How can the quality of a nearby water source be tested?*

INTRODUCTION

Background   Only 3 percent of Earth's water is fresh water, and most of it is stored in polar ice caps. Considering that humans consume great amounts of fresh water, scientists must be concerned with both the quantity of fresh water available to the regions of Earth and the quality of this precious resource. Criteria for water quality differ according to the water's use. For example, the standard of water quality for human consumption differs from that used in industry or that needed to sustain aquatic life. Contaminants in any of these types of water can be harmful.

Goals   In this investigation, you will **perform** two tests to determine the quality of a sample of water from a nearby source. You will **determine** the amount of oxygen dissolved in the water and **test** for the presence of coliform bacteria. You will then use your data to **deduce** the quality of the water from the source you have chosen, and **infer** the possible effects of the water's quality on the ecosystem.

LAB WARMUP

Concepts   Several methods are used to determine the quality of water. There are tests to detect the presence of nitrates, phosphates, minerals such as calcium and magnesium, and heavy metals such as lead. Other tests determine the pH of the water, the amount of dissolved oxygen, and the presence of coliform bacteria.

The *dissolved oxygen content (DO)* refers to the amount of oxygen gas dissolved in a sample of water. In general, a higher DO indicates relatively better water quality. The amount of oxygen that can dissolve in water depends on the temperature of the water. The lower the temperature, the more oxygen can dissolve. Dissolved oxygen content is measured in parts per million, or ppm. At room temperature (20°C) the maximum amount of oxygen that can dissolve in water is 9 parts of oxygen per 1 million parts of water (9 ppm).

The amount of bacteria present in a water source is another indicator of water quality. Coliform bacteria live in the intestines of mammals and are excreted with feces. Their presence in a water source usually suggests the water has been contaminated with fecal material that has come from pasture runoff or from untreated human sewage. The presence of coliform bacteria also suggests the water may be contaminated with other types of bacteria that may be more harmful, since many types of bacteria can thrive in the same environment. If coliform bacteria are found in a sample, further investigations can reveal the extent to which bacterial colonies are present in a given volume of water, and water safety can then be evaluated according to government standards.

Review   Section 20.2, Water Resources, and 20.3, Water Treatment, should be completed before beginning this investigation. You should also understand the following terms before you perform this investigation.

dissolved oxygen content   coliform bacteria

Make a **prediction** about the outcome of this experiment and write it in the Lab Notebook.

MATERIALS (PER GROUP)

- 4 pairs of gloves
- 4 goggles
- 4 aprons
- 1-gallon plastic bucket
- clear plastic cup
- HACH Water Quality Test Kits
- Dissolved Oxygen Test Kit
- Presence/Absence Kit for Total Coliform
- thermometer
- water bath

PROCEDURE

CAUTION: The chemicals used in the following test kits may be harmful. Do not touch any chemicals unless instructed to do so. Read all accompanying warnings carefully, and use appropriate safety equipment. Wear gloves, goggles, and an apron.

PART A DISSOLVED OXYGEN TEST

1. Pour some of the water sample into a clear plastic cup. Briefly examine the water and record your observations in the Lab Notebook. Determine the temperature of your water sample with the thermometer and record it in the Lab Notebook.

2. Fill the supplied bottle with your sample. Pour the water from the bucket into the bottle, allowing it to overflow for several seconds. Tip the bottle slightly and thrust the stopper into the bottle quickly to avoid trapping air bubbles in the bottle. It is very important not to trap any air bubbles during this procedure. If this occurs, pour out the water and repeat the procedure with more water from the bucket.

3. Remove one Dissolved Oxygen 1 Reagent Powder Pillow and one Dissolved Oxygen 2 Reagent Powder Pillow from the kit. Remove the stopper from the bottle. Open the pillows with the clippers and add the contents of each to the bottle. Carefully stopper the bottle so that no air bubbles are trapped in the bottle. Holding both the bottle and the stopper, shake vigorously to mix the contents. A brownish-orange precipitate will form if oxygen is present in the sample. Allow the sample to stand for a few minutes.

4. When the precipitate has settled so that the upper half of the solution is clear, shake the bottle again. Allow it to stand until the upper half of the solution is clear again.

5. Remove one Dissolved Oxygen 3 Reagent Powder Pillow from the kit, and open it with the clippers. Remove the stopper from the bottle, add the contents of the pillow, and carefully replace the stopper. Shake the contents to mix. If oxygen is present, the precipitate will dissolve and the solution will turn yellow.

6. Remove the plastic measuring tube and the square mixing bottle from the kit. Fill the tube with the yellow solution from step 5 and pour it into the mixing bottle.

7. Remembering to count each drop as it is added, add Sodium Thiosulfate Standard Solution drop by drop to the solution in the mixing bottle. After each drop, gently swirl the bottle to mix the contents. Continue until the solution turns colorless. Record the number of drops necessary to achieve the color change in the Lab Notebook.

8. The total number of drops added indicates the value of dissolved oxygen in parts per million. Thus, if 5 drops were used to change the solution from yellow to colorless, then the DO content is 5 ppm. Record this value in the Lab Notebook.

PART B TEST FOR THE PRESENCE OF COLIFORM BACTERIA

1. Heat a water bath to 35°C.

2. The bottle supplied for this test is presterilized. Do not contaminate the bottle by touching the inside of the cap or the neck of the bottle. Fill the bottle with 100 mL of water to be tested.

3. Using care to avoid contamination, open the vial containing the P/A broth and pour the broth into the water sample. Replace the bottle cap.

4. Place the bottle in the water bath and allow it to remain in the water bath for 24 hours. If coliform bacteria are present, the solution will change color from reddish purple to yellow. If no color change occurs after 24 hours, incubate for another 24 hours, and check again for color change. If the color of the solution still has not changed, coliform bacteria are not present in your sample.

5. Record your observations in the Lab Notebook.

Name: _____ Class: _____ Date: _____

LAB NOTEBOOK: INVESTIGATION 27

PREDICTION **A correct prediction of the water quality of students' samples would depend on their results of the tests for dissolved oxygen and presence of coliform bacteria.**

OBSERVATIONS

WATER SAMPLE

Date collected: _____

Source location: _____

DISSOLVED OXYGEN TEST

General observations about water sample: **Answers will vary, but may include cloudiness of sample, color, smell, etc.**

water temperature ____ °C **Data will vary depending on the quality of the water.**

Number of drops needed for color change _____

Dissolved oxygen (D) ____ ppm _____

COLIFORM BACTERIA PRESENCE/ABSENCE TEST

Color of solution before step 3 **Data will vary depending on the quality of the water.**

Color of solution after 24 hours _____

Color of solution after 48 hours _____

Test for coliform bacteria was (positive/negative) _____

DATA ANALYSIS

1. Most fish cannot survive at DO levels less than 4 ppm. Is there an ample amount of dissolved oxygen in your sample to sustain a fish population?
   Answers will vary; if DO>4ppm, then fish can probably survive.

2. If the water was heated, how would the amount of dissolved oxygen change?
   The amount of dissolved oxygen would decrease.

3. Why was it important not to contaminate the presterilized bottle during the test for the presence of coliform bacteria?
   Touching the inside of the cap or the bottleneck may have introduced bacteria, increasing the likelihood of a false positive result.

4. What does a positive coliform test indicate? Does this necessarily prove the water is of low quality?
   A positive result indicates the presence of bacteria but not their amount. Further investigation is needed to determine if the amount of bacteria in the sample is acceptable according to government standards.

CONCLUSION

1. **Recall** How is dissolved oxygen produced in water? Why is there a minimum level of dissolved oxygen needed for most organisms to survive?
   Dissolved oxygen is produced by photosynthetic plants in the water. Most organisms require oxygen for cellular respiration.

2. **Infer** How does your water sample rate according to the three tests you performed? What additional tests could be performed to better estimate the quality of your sample?
   Accept all logical responses. Tests for nitrates, phosphates, dissolved carbon dioxide, etc., could be performed.

3. **Integrate** If the quality of the water from the source you tested is poor, how could you improve the situation?
   Accept all logical responses. Students can express their concerns and try to effect change by writing to the appropriate political leaders, industries, environmental agencies, etc.

EXTENSION

Hypothesize How would a plant affect the quality of your water sample? Pour some of the water in a large glass jar and add a water plant such as *Elodea* or *Anacharis*. Keep the jar near a window for one day, and then repeat the tests for pH, dissolved oxygen, and coliform bacteria. Which tests showed different results? Explain your findings.

LABORATORY INVESTIGATION

28 THERMAL POLLUTION

Problem: *What effect does thermal pollution have on living things?*

INTRODUCTION

Background   Many industries and power plants produce excess heat that must be removed if production is to continue efficiently. Such heat removal is especially important in nuclear power plants. These plants produce large amounts of heat that, if not removed, could cause a dangerous meltdown. They often use circulating water carried in through pipes as a cooling system. Often, water from nearby natural sources, such as rivers and lakes, is used for this purpose. After heat exchange has taken place in a nuclear reactor, some of the heated water is used to produce steam for energy generation. Much of it, however, is returned to its source after partial cooling in a tower. This warm water raises the temperature of the body of water from which it originally came. Such an increase in water temperature as a result of human activity is called *thermal pollution*.

Goals   In this investigation, you will **observe** what happens when yeast cells are exposed to high temperatures. You will **count** the numbers of live and dead cells under various conditions, and you will **graph** the results. You will then **infer** the effect of heat on living things.

LAB WARMUP

Concepts   Temperature regulates *metabolism*, the chemical reactions occurring inside the cells of organisms. Temperature changes that are severe enough can make it impossible for these life processes to continue. It is thereby possible for organisms to die from exposure to excess heat or excess cold.

Even if the temperature increase is not great enough to cause death, it can have other harmful effects. For example, increased temperature raises rates of respiration. That increases the demand for oxygen. Under these conditions it can become difficult for organisms to obtain enough oxygen. The problem is made worse by the fact that increasing temperature decreases the amount of oxygen gas in water. This further reduces the oxygen supply. Fishes are among the organisms that can be harmed by thermal pollution. They may die as a result of increased water temperature, which raises their respiration rate and reduces the amount of dissolved oxygen.

Review   Section 21.3, Radioactivity and Thermal Pollution, should be completed before beginning this investigation. You should also understand the following terms before you perform this investigation.

thermal pollution   metabolism

Make a **prediction** about the outcome of this experiment and write it in the Lab Notebook.

MATERIALS (PER GROUP)

- 7 test tubes
- grease pencil
- 10 mL water
- test-tube rack
- 12 dry yeast granules
- protective gloves
- 250-mL beaker
- hot plate
- test-tube holder
- clock or watch that indicates seconds
- dropper
- glass slide
- methylene blue stain
- toothpick
- coverslip
- microscope

PROCEDURE

1. Label seven test tubes with the numbers *0, 10, 20, 30, 40, 50,* and *60*. Pour 4 mL of tap water into each test tube. Place the tubes in the test-tube rack as shown in Figure 28.1.

2. Drop 2 granules of dry yeast into each test tube. Gently shake the tubes to dissolve the yeast.

3. **CAUTION: Be careful not to burn yourself with the hot water and heat source. Wear goggles, protective gloves, and a lab apron.** Fill a beaker roughly three-fourths full with tap water. Heat the beaker over a hot plate to boil the water.

4. Using a test-tube holder, hold the tube labeled *10* in the boiling water for 10 seconds. Remove it and return it to the rack.

5. Repeat step 4, holding the test tubes labeled *20, 30, 40, 50,* and *60* in the boiling water for 20 seconds, 30 seconds, 40 seconds, 50 seconds, and 60 seconds, respectively. Let all the tubes cool for several minutes. Do not heat the tube labeled *0*.

6. Gently shake tube 0, and use a dropper to transfer 1 drop of its contents onto a microscope slide. Add 1 drop of methylene blue stain to the drop of yeast, and use a toothpick to mix the two. Place a coverslip on top. **CAUTION: Be careful not to spill the methylene blue, as it can stain your skin and clothing.**

7. Observe the slide under a microscope at high power. Live yeast cells will be pale blue in color. Dead yeast cells will be a darker blue. Counting from the top of the field of vision, and moving downward, observe 100 cells. In the table in the Lab Notebook, record the number of cells that are alive and the number that are dead. If there are fewer than 100 cells in the field, move to another area and continue counting until you reach 100. Record the percentage of live cells.

8. Repeat steps 6 and 7 for each of the other 6 test tubes. Reuse the slide and coverslip. rinsing it before each use.

9. Use the grid in the Lab Notebook to graph the percentage of live cells versus the time in hot water.

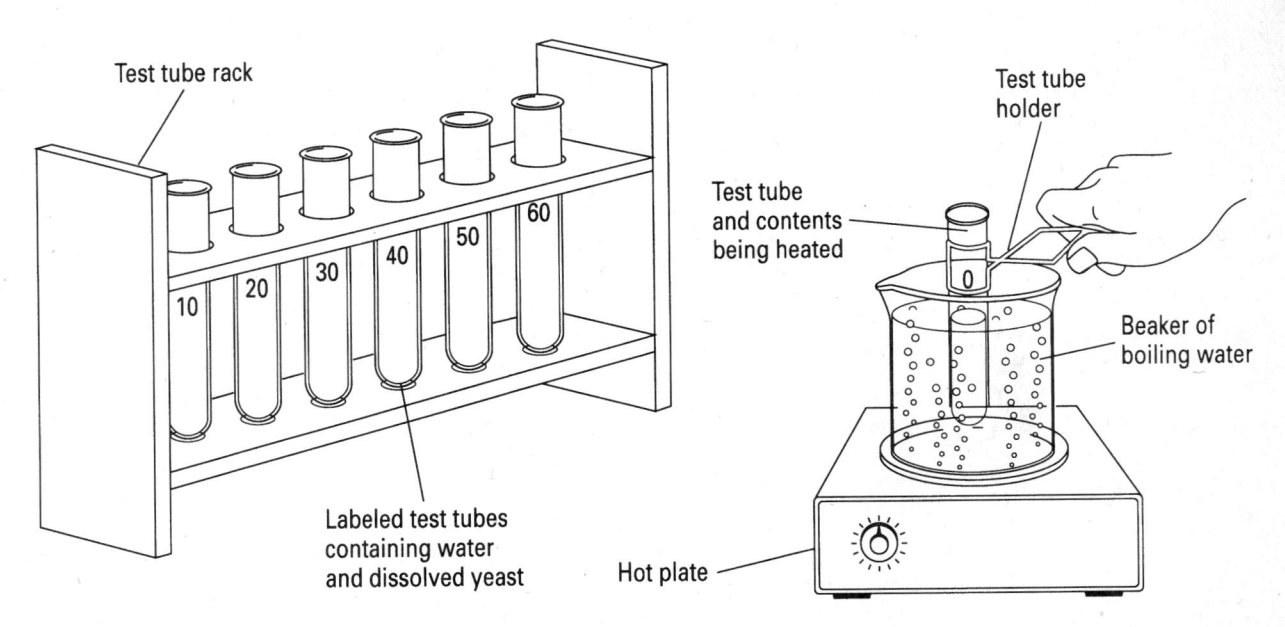

Figure 28.1 Setup of test-tube rack and hot plate

Name: _____ Class: _____ Date: _____

LAB NOTEBOOK: INVESTIGATION 28

PREDICTION **A correct prediction is that the yeast cells will die when exposed to high temperatures.**

OBSERVATIONS

NUMBERS OF LIVE AND DEAD YEAST CELLS

Time in hot water (sec)	Number of live cells	Number of dead cells	Percentage of live cells
0			
10	**The number of live yeast cells will decrease as length of exposure increases.**		
20			
30			
40			
50			
60			

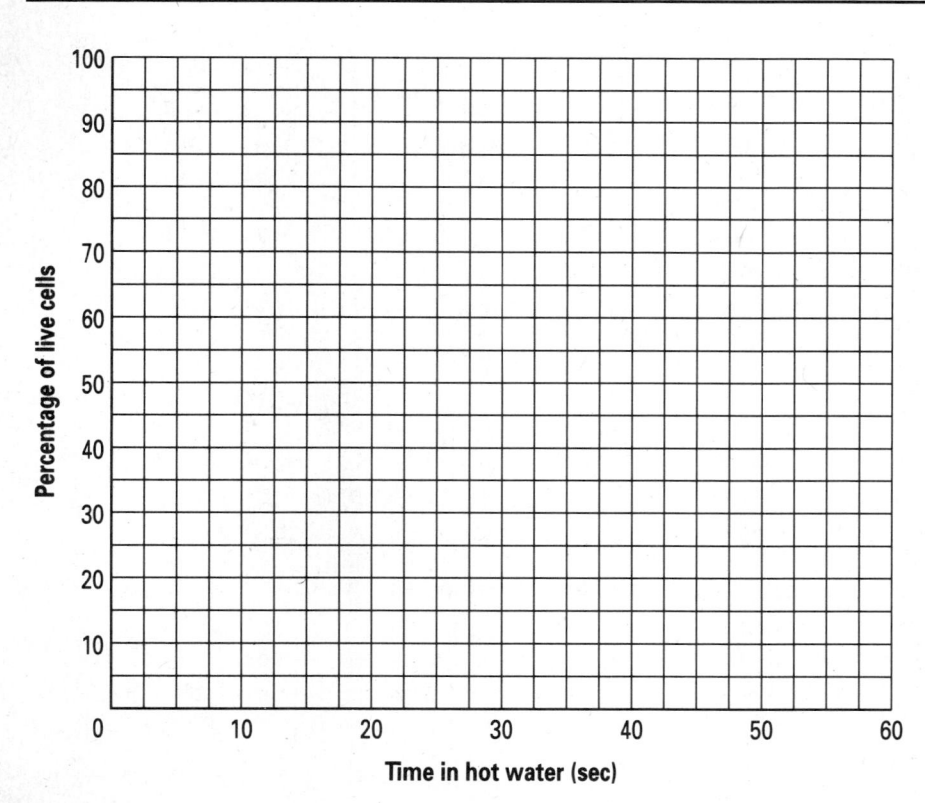

[Graph: Percentage of live cells (y-axis, 0–100) vs. Time in hot water (sec) (x-axis, 0–60)]

DATA ANALYSIS

1. Given that 100 cells were observed in each case, how does the number of live cells relate to the *percentage* of live cells?
   The two values are numerically equal.

2. What relationship do you notice between the percentage of live cells and the amount of time the yeast remained in hot water?
   The percentage of live cells decreases as the amount of time in hot water increases.

3. How well does your count represent the ratio of living to dead cells in each tube? How could you make the ratio more accurate?
   The ratio is probably representative. A larger sample size would make the ratio more accurate.

CONCLUSION

1. **Infer** Explain the effect of the hot water on the yeast cells.
   The temperature of the water bath was so high that the life processes of yeast cells could not be maintained. The longer the test tubes were in the water, the higher the temperature rose and the more yeast cells died.

2. **Generalize** Relate your observations to the effects of thermal pollution on living things.
   Like the hot plate, thermal pollution raises water temperature to a point that is harmful to the metabolism of organisms. If the temperature is high enough, it will kill the organisms.

3. **Infer** Suppose thermal pollution in a pond was not serious enough to directly kill or harm organisms. Why might some of the organisms still be adversely affected?
   The larger organisms might suffer from the decreased oxygen content of the water or from changes in the plankton community.

4. **Infer** What could concerned citizens do to help reduce or limit thermal pollution in a nearby water source?
   Accept all logical responses. Students may suggest monitoring industrial operations, requiring the use of cooling towers, limiting the access of industries to bodies of water, etc.

EXTENSION

Research Assess the extent and effects of thermal pollution in your own area by arranging interviews with local environmental groups and representatives of industries. If possible, try to make on-site inspections, with the permission of property owners and in the company of an adult. Write a report on your findings.

LABORATORY INVESTIGATION

29 MODELING A WET SCRUBBER

Problem: *How does a wet scrubber operate, and how can a working model of a wet scrubber be constructed?*

INTRODUCTION

Background   Air pollution causes many serious problems. For example, it gives rise to acid rain, which can kill trees and make bodies of water uninhabitable by organisms. Air pollution also damages buildings, causes health problems in humans, and contributes to global warming. Many industries and power plants produce large quantities of pollutants. These include particulates and also poisonous gases, such as sulfur dioxide. When released into the atmosphere, these industrial wastes can produce serious levels of air pollution. Controlling industrial emissions is thus one way of reducing air pollution problems. One common method of removing pollutants from exhaust is by exposing it to water in a device called a wet scrubber.

Goals   In this investigation, you will **model** a wet scrubber, **observe** it in operation, and **infer** its usefulness in reducing air pollution.

LAB WARMUP

Concepts   A *wet scrubber* is a simple pollution-control device. It is used in many industries to reduce harmful emissions. Air containing harmful substances produced by industrial processes is passed into the wet scrubber. The scrubber contains water and, usually, absorbent material suspended in the water. The polluted air enters the water mixture through a tubelike device called an impinger. The impinger has a narrow opening that produces small bubbles of gas. The pollutants dissolve or settle in the water mixture. Often they react with the absorbent materials. For example, compounds of calcium or sodium may be dissolved in the water. They react with sulfur dioxide gas emissions to produce solid compounds that are then collected. The solids may then be disposed of or processed to make useful compounds.

Review   Section 22.4, Controlling Air Pollution, should be completed before beginning this investigation. You should also understand the following terms before you perform this investigation.

pollution   wet scrubber   impinger

Make a **prediction** about the outcome of this experiment and write it in the Lab Notebook.

MATERIALS (PER GROUP)

- goggles
- aprons
- 3 2-hole rubber stoppers
- 3 5-cm pieces of glass tubing
- 2 impingers (15-cm pieces of glass tubing with a smaller opening at one end)
- 15-cm piece of glass tubing
- paper towel
- 3 500-mL flasks
- water
- 3 40-cm lengths of rubber tubing
- ringstand
- beaker clamp
- wire gauze
- Bunsen burner
- marker
- vacuum source
- heat-resistant gloves
- matches

PROCEDURE

1. Put on your goggles and apron.
2. Fit each of 2 two-hole rubber stoppers with a short piece of glass tubing and an impinger. Fit a third two-hole stopper with another short piece of glass tubing and a longer piece of glass tubing that has not been tapered. **CAUTION: Be careful when handling glass objects and when fitting glass tubing into stoppers.** Follow the figure on this page.
3. Place a crumpled paper towel into a 500-mL flask. Fill each of two other 500-mL flasks about $2/3$ full of water. Fit the stoppers into the flasks as shown in Figure 29.1.
4. Connect the pieces of glass tubing, as shown, with lengths of rubber tubing. One length of rubber tubing should lead to a vacuum source set up by your teacher. Do not turn on the vacuum source yet.
5. Set up a ringstand, clamp, and wire gauze as shown, and place on it the flask containing the paper towel. Put a Bunsen burner under the flask. Make sure your setup is like that shown in the figure.
6. Label the three flasks *1*, *2*, and *3* as shown.
7. Light the burner and gently heat the flask above it. **CAUTION: Be careful when using the burner and handling hot objects. Use heat-resistant gloves when touching hot objects.** When smoke first appears inside the heated flask, turn on the vacuum to draw a stream of smoke through the flasks of water. Record your observations in the Lab Notebook. After about 5 minutes, turn off the burner.
8. Examine the flasks of water and record your observations in the Lab Notebook.
9. When the apparatus has cooled, disassemble it and dispose of contaminated water as directed by your teacher.

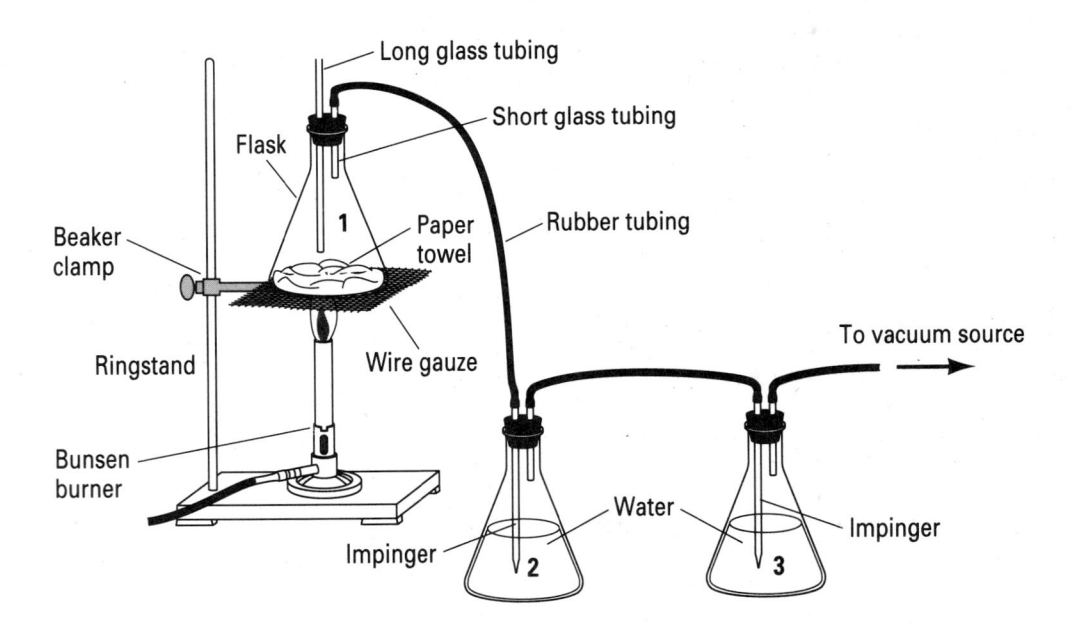

Figure 29.1 Setup of Bunsen burner and flasks

Name: _____ Class: _____ Date: _____

LAB NOTEBOOK: INVESTIGATION 29

PREDICTION **A correct prediction is that the water in the wet scrubber will remove pollutants from the air.**

OBSERVATIONS

APPEARANCE OF FLASK CONTENTS

Flask 1
Smoke in flask.

Flask 2
Little or no smoke in flask; water discolored.

Flask 3
No smoke in flask; water clear.

DATA ANALYSIS

1. Why did the water in Flask 1 change in appearance?
   The water became discolored because materials from the smoke were dissolved or suspended in the water.

2. Why was an impinger used rather than an untapered piece of glass tubing?
   The impinger breaks the smoke up into smaller bubbles. Smaller bubbles have more surface area per unit volume than larger bubbles, and so expose more smoke to the water. More materials are therefore removed.

3. Do you think the wet scrubber was able to remove all the pollutants in the air? Explain your answer.
   It is unlikely that all the pollutants were removed. The process was probably less than 100% efficient, even for soluble materials. Also, some colorless insoluble gases may have been produced and passed out of the water into the vacuum source.

CONCLUSION

1. **Generalize** Based on your observations, explain why wet scrubbers are useful in reducing air pollution.
   Accept all logical responses. The water in the scrubbers dissolves or traps pollutants and prevents them from escaping into the atmosphere.

2. **Model** How would you have to adapt the basic design of your wet scrubber to make one that would be suitable for industrial use?
   Accept all logical responses. The apparatus would have to be much larger and would use much more water. More stages for pollutant removal would be needed, construction materials would have to be sturdier, etc.

3. **Reevaluate** Based on your observations, what new potential pollution problem is created by the use of wet scrubbers? How should this problem be handled?
   The contaminated water that is produced must be disposed of safely. The water must be treated to remove pollutants or must be disposed of in a way that would avoid environmental contamination.

EXTENSION

Model Design a system that would treat polluted water from air scrubbers. The system should remove particles, dissolved solids, and gases from the water. Does your system completely solve the pollution problem?

LABORATORY INVESTIGATION

30 ACID RAIN AND SEED GROWTH

Problem: *How does acid rain affect seed germination?*

INTRODUCTION

Background Industries and motor vehicles produce gaseous oxides of nitrogen and sulfur. For example nitrogen and oxygen in the air can combine under high-temperature engine conditions to produce nitrogen dioxide (NO_2). The equation for the reaction is: $N_2 + 2O_2 \rightarrow 2NO_2$. Sulfides in fuels can combine with oxygen to make sulfur dioxide (SO_2) and sulfur trioxide (SO_3). Such oxides combine with water in the atmosphere to make acids. For example, nitrogen dioxide and sulfur trioxide combine with water. They form nitric acid (HNO_3) and sulfuric acid (H_2SO_4), respectively. The equations for these reactions are: $3NO_2 + H_2O \rightarrow 2HNO_3 + NO$ and $SO_3 + H_2O \rightarrow H_2SO_4$. The presence of these acids causes rain to be acidic. Acid rain damages trees, crops, and buildings. It can make lakes so acidic that fish cannot survive.

Goals In this investigation, you will moisten seeds with acid solutions of various concentrations. You will then **observe** how many seeds germinate in each case. You will **graph** the information and **infer** the effect of acidity on seed germination.

LAB WARMUP

Concepts The acidity of solutions is measured using the pH scale, which extends from 1 to 14. The pH of a solution is defined as the negative of the logarithm of the hydrogen-ion concentration ($-\log [H^+]$). For example, $[H^+]$ in pure water is 1.0×10^{-7}. The pH of pure water is therefore $-\log (1.0 \times 10^{-7})$, or 7. The pH of acidic solutions is less than 7, and that of basic solutions is greater than 7. The lower the pH below 7, the more acidic is the solution.

Paper that has been soaked in a mixture of indicator solutions is called pH paper. It turns different colors, from red (highly acidic), through orange, yellow, yellow-green (neutral), and green, to blue (highly basic). Using pH paper is a quick and simple way to estimate the pH of a solution.

Review Section 22.3, Global Effects of Air Pollution, should be completed before beginning this investigation. You should also understand the following terms before you perform this investigation.

acid   germinate   pH

Make a **prediction** about the outcome of this experiment and write it in the Lab Notebook.

MATERIALS (PER GROUP)

- rubber gloves
- 25 radish or alfalfa seeds
- small beaker
- mold inhibitor
- container of solution of unknown pH, provided by your teacher
- litmus paper
- water
- paper towels
- plastic bag
- dropper
- twist tie
- marker

© *Addison-Wesley Publishing Company, Inc. All Rights Reserved.*

PROCEDURE

1. **CAUTION: Mold inhibitor is harmful to skin and can damage clothing. Wear goggles, rubber gloves, and an apron throughout the investigation.** Place 25 seeds into a small beaker and add just enough mold inhibitor to cover them. The inhibitor will help prevent the seeds from rotting later on.

2. **CAUTION: Be very careful handling acid solutions. Report any spills to your teacher.** Let the seeds soak for 10 minutes. While the seeds are soaking, obtain a container of solution from your teacher. The solution will have a pH from 1 to 7. Determine the approximate pH of your solution by using a piece of pH paper. Compare the color of the paper to that shown on the scale provided with the paper. Choose the number that corresponds best, and record this information in the Lab Notebook.

3. Drain the mold inhibitor from the seeds and rinse the seeds with fresh tap water. Place the seeds on clean paper towels, and blot them dry. Dispose of the inhibitor as directed by your teacher.

4. Place three paper towels into a plastic bag, folding the towels in half. Place the 25 seeds separately between the layers of paper towels, as shown in Figure 30.1. Use a dropper to moisten the towels thoroughly with your assigned solution, without soaking them. Use a twist tie to close the bag. Label it and put it in a moderately warm place.

5. A few days later, open the bag and moisten the seeds once again with the same solution. **CAUTION: Remember to wear goggles, gloves, and an apron.** Notice whether any seed coats have split or if any seeds are germinating.

6. One week after the beginning of the investigation, open the bag and examine the seeds. Count the number of seeds that have germinated. Record this number in the Lab Notebook. When directed to do so by your teacher, also record this number on the chalkboard, next to the pH that corresponds to that of your assigned solution.

7. In the Lab Notebook, copy down the germination data written on the chalkboard by all the students. Notice that each pH value was assigned to at least two groups of students. Average the seed germination numbers for each pH value and record the averages in the Lab Notebook. Dispose of the seeds and acid solutions as directed by your teacher.

8. Make a bar graph of the average number of seeds germinated for each pH in the Lab Notebook.

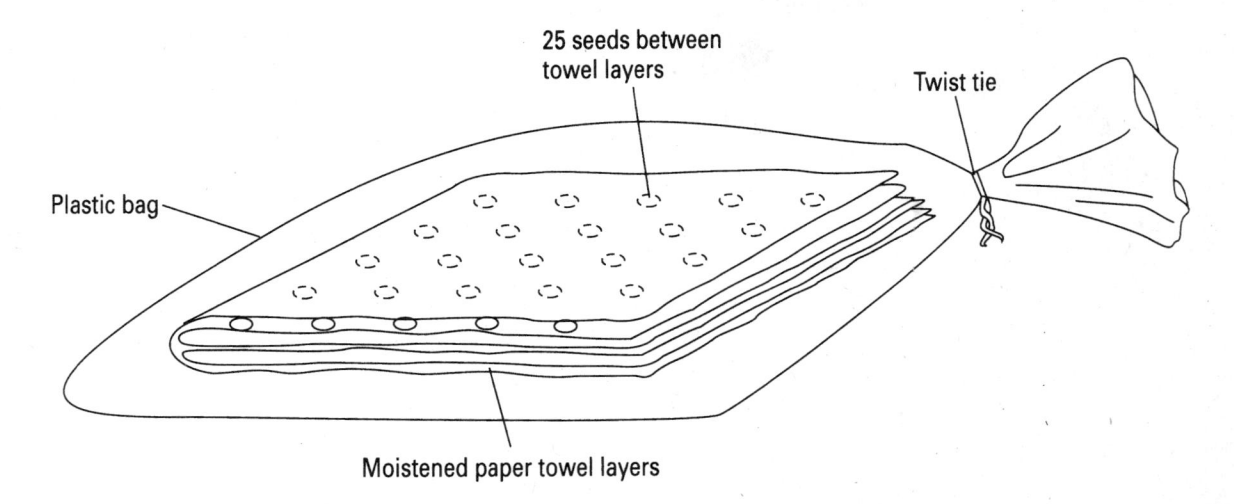

Figure 30.1 Setup of towel layers

Name: _____ Class: _____ Date: _____

LAB NOTEBOOK: INVESTIGATION 30

PREDICTION *A correct prediction is that the presence of acid decreases the number of seeds that germinate*

OBSERVATIONS

pH of solution used: _____

Number of seeds germinated within 1 week: _____

CLASS DATA ON RATE OF SEED GERMINATION

pH	Number of seeds germinated		
	First group	Second group	Average
1			
2	**Class data will vary somewhat, depending on the seeds used and conditions**		
3	**such as temperature. However, the number of seeds that germinate in more**		
4	**acidic solutions should be lower than the number of seeds that germinate in**		
5	**less acidic solutions or in pure water.**		
6			
7			

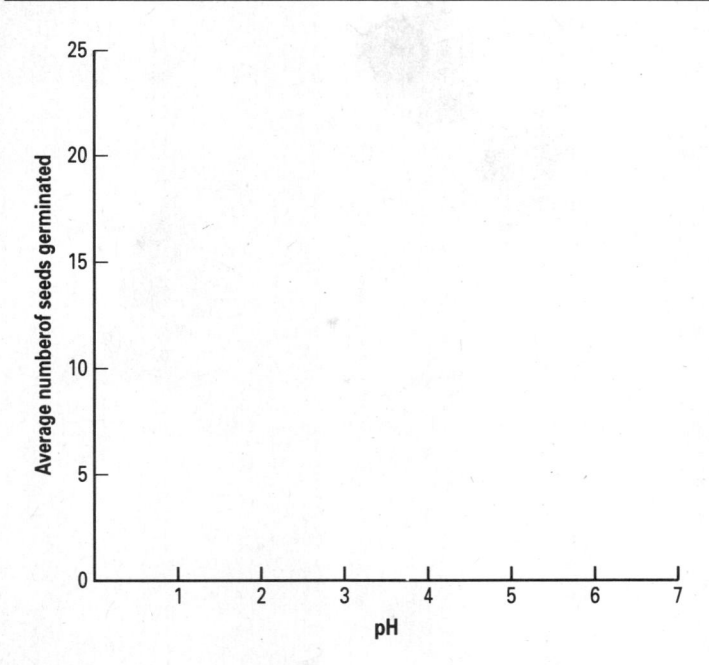

DATA ANALYSIS

1. What do you notice about the relationship between seed germination and pH?
   As pH decreases, the number of seeds that germinate decreases.

2. Why were data gathered using pure water (pH 7) as well as acidified water?
   The use of pure water provided a control that provided information on the viability of the seeds for comparison with the experimental seeds. The control showed how many seeds would be unlikely to sprout even under ideal conditions.

3. Between what two pH values do you notice the most dramatic change in the number of seeds that germinated?
   Answers will vary, but the most dramatic change should occur between either pH 4 and pH 3, or pH 3 and pH 2. There should be little or no germination at pH 1.

CONCLUSION

1. **Generalize** Given what you have observed, what is the likely effect of acid rain on seed germination and plant reproduction? How acidic must rain be to produce any significant effect?
   Acid rain reduces the successful reproduction of plants. A significant effect may occur as high as pH 5, but the effect becomes dramatic at lower pH values.

2. **Infer** Some rain has been found to be as acidic as vinegar (pH approximately 2.8). Do you think this rain has affected seed germination? Explain your answer in terms of your observations in this investigation.
   Given that pH values below 3 produce much lower rates of germination, it is likely that rain at such a low pH level has had a significant effect on seed germination.

3. **Infer** Do you think plants that are able to germinate despite acidic conditions cannot be harmed by acid rain once they have sprouted? Explain your answer.
   The acid may continue to affect the seedlings adversely, preventing them from growing into adults. It may also kill adult plants prematurely or damage them and affect their ability to reproduce.

EXTENSION

Predict Does acid rain affect the germination of all seeds to the same extent? Find out by repeating the investigation using different kinds of seeds. Carry out the experiment under your teacher's supervision and observe all safety precautions.

LABORATORY INVESTIGATION

31 GLOBAL WARMING AND BIODIVERSITY

Problem: *What effects can global warming have on living things?*

INTRODUCTION

Background   Certain atmospheric gases, such as carbon dioxide and methane are called greenhouse gases. Such gases allow sunlight to reach Earth. However, they prevent the infrared radiation that is then produced from returning to space. This natural trapping of infrared rays is called the greenhouse effect. The greenhouse effect helps warm Earth and its atmosphere.

Some greenhouse gases, such as carbon dioxide, have been building up quickly in Earth's atmosphere over the past several decades. A number of direct studies on the atmosphere have shown this. Other studies have shown that the concentration of carbon dioxide has been building up in sea water as well. The sea water absorbs some of the excess carbon dioxide from the atmosphere.

Extra heat is being trapped by these gases in the atmosphere. This human-caused increase in temperature is called global warming. There is evidence that climates around the world may be changing as a result of global warming.

Goals   In this investigation, you will **infer** what will happen to various kinds of organisms as conditions in their habitat change. You will then **predict** which kinds of organisms will survive and which will become extinct.

LAB WARMUP

Concepts   Different kinds of organisms occupy specific ecological niches. Certain climatic conditions exist in these niches. Because of temperature increases associated with global warming, the habitats may no longer support organisms that once lived in them. Certain other abiotic factors, such as the quantity of rainfall, may also change as a result of global warming. As these factors change, some of the organisms not adapted to the new conditions will migrate. Others will perish. The result can be a widespread habitat destruction and loss of biodiversity.

Review   Section 22.3, Global Effects of Air Pollution, and Section 23.1, The Loss of Biodiversity, should be completed before beginning this investigation. You should also understand the following terms before you perform this investigation.

habitat   extinction   biodiversity   abiotic factors   migration

Make a **prediction** about the outcome of this experiment and write it in the Lab Notebook.

MATERIALS (PER GROUP)

- colored pencil

PROCEDURE

1. Imagine an island made up of two climate zones, as in Figure 31.1. The northern half of the island has an average yearly temperature of 15°C and an average yearly precipitation of 50 cm. The climate values for the southern half are 20°C and 40 cm. A nature reserve occupies the lower part of the southern half of the island. Eight kinds of organisms found nowhere else on Earth live within the reserve. To survive, each requires that the average temperature remain below a certain maximum and that average rainfall remain above a certain minimum. No organism confined within the reserve can migrate off the island. The chart below provides detailed information about each kind of organism in the reserve.

REQUIREMENTS FOR ORGANISMS IN THE RESERVE

Organism	Max. temp. (°C)	Minimum rainfall (cm)	Max. migration per year (km)	Possible food sources
yellow finch	25	30	200	windseed, velvet grass
speckled frog	25	35	1	red beetle
furry rat	30	25	25	windseed, velvet grass
golden panther	25	40	100	furry rat
brown coyote	30	30	50	speckled frog
red beetle	20	30	10	velvet grass
windseed	25	30	3	—
velvet grass	20	30	1	—

2. Now imagine that global warming is taking place at a rapid rate. Scientists determine that over the next 10 years, the average temperature in the southern half of the island will rise by 0.5°C per year, to a value of 25°C. Average rainfall will decrease by 1 cm per year to a value of 30 cm. In the northern half of the island, the average temperature will be 20°C and the average rainfall will be 40°C at the end of the 10-year period. Note that these values are the same as the original values for the southern half of the island.

3. The organisms in the nature reserve are now permitted to migrate in order to save themselves from extinction. Consider what will happen to the habitat range of each species. Think in terms of possible migration, changes in food supply, and ultimate survival or extinction. The original habitats of the organisms are shown on the maps in the Lab Notebook as darkened areas. The distance between each pair of horizontal lines represents 50 km, as shown in the key. The heavy line across the center represents the climate boundary between the northern and southern halves of the island. Notice that only one of the organisms—windseed—is originally present on the northern half of the island as well as in part of the reserve. Use the maps in your Lab Notebook and the chart above to answer the questions in the Data Analysis section.

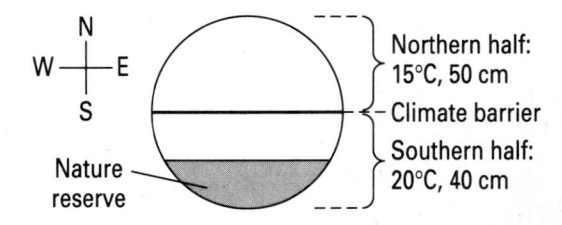

Figure 31.1 Climate zones

Name: _____ Class: _____ Date: _____

LAB NOTEBOOK: INVESTIGATION 31

PREDICTION   **A correct prediction is that some kinds of organisms will survive, in some cases by migrating, and that some will become extinct, either from changes in abiotic conditions or from the disappearance of food sources.**

OBSERVATIONS

MAPS OF HABITATS OF ORGANISMS

Organism	Original habitat	Habitat in 10 years
Yellow finch	Northern half clear, southern half shaded	Northern half shaded, southern half clear
Speckled frog	Both halves clear	Both halves clear
Furry rat	Northern half clear, southern half shaded	Both halves shaded
Golden panther	Northern half clear, southern half shaded	Both halves shaded
Brown coyote	Northern half clear, southern half shaded	Both halves clear
Red beetle	Both halves clear	Both halves clear
Windseed	Both halves shaded	Both halves shaded
Velvet grass	Northern half clear, southern half shaded	Northern half clear, southern half shaded

Key: 50 km; Northern half (clear), Climate boundary (solid line), Southern half (reserve, shaded)

131

DATA ANALYSIS

1. State whether each species could survive under the new abiotic conditions in the original reserve area (assume that food is not an issue). If a species cannot survive the new temperature or rainfall conditions, state which of these conditions you think the species would find intolerable.

   a. yellow finch __yes__
   b. speckled frog __no, insufficient rain__
   c. furry rat __yes__
   d. golden panther __no, insufficient rain__
   e. brown coyote __yes__
   f. red beetle __no, excessive temperature__
   g. windseed __yes__
   h. velvet grass __no, excessive temperature__

2. Assess each species' ability to migrate quickly enough to reach a habitat where climatic conditions could support it (assume food is not an issue).

   a. yellow finch __yes__
   b. speckled frog __no__
   c. furry rat __yes__
   d. golden panther __yes__
   e. brown coyote __yes__
   f. red beetle __no__
   g. windseed __yes__
   h. velvet grass __no__

3. State whether each species would have a food source anywhere within the region to which it would migrate. Briefly explain each answer.

   a. yellow finch __Yes; there would be windseed.__
   b. speckled frog __No; there would be no red beetles.__
   c. furry rat __Yes; there would be windseed.__
   d. golden panther __Yes; there would be furry rats.__
   e. brown coyote __No; there would be no speckled frogs.__
   f. red beetle __No; there would be no velvet grass.__
   g. windseed __No food source would be needed (this is a plant).__
   h. velvet grass __No food source would be needed (this is a plant).__

4. Use your answers to the questions above to determine the habitat, if any, of each kind of organism after 10 years. In order to survive to this point, an organism would have had to be able to reach areas with the right climate and food source for its survival. Show the locations of these final habitats by coloring in the appropriate areas on the maps on the right in the Lab Notebook.

CONCLUSION

1. **Analyze** Which kinds of organisms on the island will survive the change? Which will become extinct? What will happen to the biodiversity of the island?

   Yellow finch, furry rat, golden panther, and windseed will survive. Speckled frog, brown coyote, red beetle, and velvet grass will become extinct. Biodiversity will decrease.

2. **Infer** Given what you have learned from this investigation, why are changes that might result from global warming dangerous to individual species and to life as a whole on Earth?

   Individual species might become extinct because of inabilities to tolerate changes in abiotic conditions, inabilities to migrate quickly enough or loss of food source.

EXTENSION

Research Conduct library research to find out more about global warming and evidence that global warming may be taking place. Also investigate changes that may have already been caused by such warming and changes that may occur. Write a report on your findings.

LABORATORY INVESTIGATION

32 COMPOSTING

Problem: *How is compost made? What are the benefits of composting?*

INTRODUCTION

Background Food wastes and yard wastes from your household do not have to be disposed of in a landfill. They can be composted instead. A *compost pile* is a mass of decaying organic matter. Potato peelings, eggshells, onion skins, grass clippings, leaves—practically every organic material except meats and fats—can be added to a compost pile. These organic materials decompose and change into a nutrient-rich, organic fertilizer. Organic materials that decompose readily in a compost pile do not decompose easily in a landfill because a landfill lacks one of the essential ingredients for decomposition—air. When these materials get trapped on the bottom of the landfill, bacteria and other organisms cannot break them down.

Contrary to what people may think, a properly maintained compost pile does not smell bad and does not attract animals. And, properly placed, it need not be unsightly. Maintaining the compost pile takes surprisingly little effort.

Goals In this investigation, you will **construct** a compost pile and monitor it for three weeks. You will **measure** the temperature of the pile, and **observe** other features of the decomposition process. You will then **evaluate** the effectiveness of composting at reducing the amount of garbage disposed in landfills.

LAB WARMUP

Concepts The conversion of wastes into compost through decomposition requires five essential ingredients:

1. **Organic material to be decomposed** This supplies the carbon essential for the decomposition process and the minerals that make nutrient-rich compost.

2. **Decomposers** These can be found in garden soil and freshly pulled weeds (many microorganisms are found on the roots of weeds).

3. **Nitrogen** Some organic material is higher in nitrogen content than others. A 30:1 ratio of carbon to nitrogen is ideal for a compost pile. Dried leaves, wood chips and sawdust are high in carbon, ranging in carbon/nitrogen ratios from 150:1 to 500:1. Paper has a carbon/nitrogen ratio of 170:1. Vegetable and fruit peelings range from 12:1 to 35:1. Grass clippings are higher in nitrogen (15:1). Materials rich in nitrogen include alfalfa, clover, leather, dust, nut shells (especially peanut), and hair. In targeting a carbon/nitrogen ratio for you pile, estimate the amounts of materials by weight, not volume. Too much nitrogen can result in the production of ammonia, which has a bad odor. Too little nitrogen reduces the speed of the decomposition process, since nitrogen is needed for bacteria to break down organic matter.

4. **Water** The heap should be kept moist but not soggy. It should feel like a squeezed-out sponge. If the pile is too soggy, the materials may become matted, preventing proper aeration.

5. **Air** Although the decomposing microorganisms can survive without air, they will change to anaerobic cellular respiration. Anaerobic respiration slows the decomposition process, producing foul-smelling hydrogen sulfide gas.

Cellular respiration produces heat. Thus, the internal temperature of the pile varies with the amount of decomposition taking place inside. A slight drop in temperature indicates that the bacteria have reduced their activity due to lack of oxygen, and the pile should be turned to aerate it.

Do not add the following materials to your compost pile: meats and meat products, vegetable and mineral oils, fats and grease, plastic, or synthetic fibers such as polyester or nylon. These materials either slow down decomposition or will not decompose. Also, since some inks can be toxic, do not use paper with ink if you intend to add the compost to vegetable gardens.

© *Addison-Wesley Publishing Company, Inc. All Rights Reserved.*

Review Section 19.4, Controlling Land Pollution, and Section 24.2, Recycling, should be completed before beginning this investigation. You should also understand the following terms before you perform this investigation.

compost   decomposition   cellular respiration   aeration   anaerobic

Make a **prediction** about the outcome of this experiment and write it in the Lab Notebook.

MATERIALS (PER CLASS)

- wire or screen compost bin
- organic yard and food wastes
- dirt (or nonsterile soil)
- additional nitrogen source, if needed
- water
- watering can or hose
- 10 to 20 earthworms
- thermometer
- pitchfork

PROCEDURE

1. Select a site for the compost pile. Avoid areas of direct sunlight and strong winds. Assemble the compost bin.
2. Gather fruit and vegetable wastes, leaves, grass clippings, eggshells (rinsed with water), sawdust, and other materials for your compost pile. Remember to try to keep a carbon-to-nitrogen ratio of about 30:1. Shred all the materials that are not already in very small pieces.
3. Pour a thin layer (about 2 cm) of dirt into the compost bin. Alternate layers of high-nitrogen materials and low-nitrogen materials (about 8-cm thick each) and dirt until the materials are used up or until the bin is full.
4. Sprinkle water over the pile with a hose or watering can. Do not make the pile soggy; it should feel like a squeezed-out sponge.
5. Add the earthworms to the dirt at the top of the pile.
6. Insert the thermometer into the pile. Record the internal temperature of the pile (in degrees Celsius) in your Lab Notebook.
7. Every day for the next 3 weeks, record the internal temperature of the pile, the texture of the compost, the odor of the compost, and any observed changes in the materials placed in the compost pile. Water the compost pile occasionally to keep it moist.
8. Aerate the pile every 3 to 4 days or whenever there is a slight drop in the internal temperature. When turning the pile, the outside materials should end up on the inside, and the materials on the bottom should end up on the side or top of the pile. Record in the Lab Notebook each time you aerate the pile.
9. On a separate sheet of graph paper plot the internal temperature of the pile over the 3-week period. Label the numbers of days on the *x*-axis and temperatures on the *y*-axis. Indicate when you aerated the pile on the graph.

Name: _____ Class: _____ Date: _____

LAB NOTEBOOK: INVESTIGATION 32

PREDICTION **A correct prediction is that organic wastes in a compost pile will decompose over time and form a nutrient-rich soil.**

OBSERVATIONS

OBSERVATIONS OF COMPOST PILE

Day	Date	Temp (°C)	Aerate	Texture	Odor	Other observations
1						
2						
3						
4		Data will vary but should show a change in temperature and				
5		texture over time.				
6						
7						
8						
9						
10						
11						
12						
13						
14						
15						
16						
17						
18						
19						
20						
21						

DATA ANALYSIS

1. Why did the materials for the compost pile need to be shredded?
   Smaller pieces increase the surface area of the material, allowing microorganisms to decompose the material more quickly.

2. Did your compost pile ever have a foul smell? If so, what might have caused this?
   If the pile smelled bad, it was probably due to incomplete or infrequent aeration of the pile, causing the microorganisms to undergo anaerobic respiration. It could also be due to an excess of nitrogen in the pile, producing ammonia.

3. Did all the materials completely decompose after three weeks? If not, which materials were still identifiable? Why didn't they decompose?
   Answers will vary. If any identifiable materials remain, they probably were not shredded into small enough pieces. Wood chips or sawdust may not have had sufficient time to decompose.

4. How did the internal temperature of the compost change over the course of the three weeks?
   Answers will vary, but the temperature should have increased, decreased slightly before aeration, and decreased once the materials were completely decomposed.

CONCLUSION

1. **Infer** Why does the internal temperature of the compost pile increase? Why does it decrease?
   When microorganisms decompose the material during cellular respiration, heat is given off; the temperature decreases when there is not enough oxygen to sustain aerobic cellular respiration or when there is no more organic material left to decompose.

2. **Infer** What happened to the organic wastes that were decomposed in the compost pile?
   The wastes were broken down into smaller molecules.

3. **Integrate** What are the benefits of having a backyard compost pile?
   Accept all logical responses. Reduction of landfill trash and use of compost for gardening should be included.

EXTENSION

Research Contact the waste management system in your town, city or community. Find out what percentage of landfill space is taken up by yard waste. Devise a plan for an alternative disposal method for yard waste that includes composting. Your plan should be practical and produce enough income to pay for itself.

LABORATORY INVESTIGATION

33 HAZARDOUS WASTES SURVEY

Problem: *How many hazardous materials are used within your school grounds? How should hazardous wastes be disposed of? What are the alternatives to the use of hazardous substances?*

INTRODUCTION

Background   You may not realize how many hazardous substances are in your home and school. Batteries, many cleaning solutions, insecticides, weed killers, solutions for copiers and ditto machines, and the chemicals in your science laboratory are all considered hazardous by the Environmental Protection Agency (EPA). More than 236 million tons of hazardous wastes are produced in the United States each year. Many of these wastes are handled and disposed improperly. For example, if a container of a hazardous material is thrown in the trash, it will be disposed in a landfill. Over time the container will probably leak, and, with the help of rainfall, the hazardous material can seep into the ground and into groundwater. Dumping hazardous wastes into lakes and rivers pollutes the water and harms aquatic life. Improper burning of hazardous wastes causes air pollution.

Solutions to the problem of hazardous waste disposal are not simple. Once the waste is generated, proper handling and disposal are expensive. The best way to solve the hazardous waste problem is not to buy products containing hazardous substances in the first place, thereby eliminating the need for costly handling and disposal. Many safer alternatives to hazardous products are available. However, a safe alternative cannot be found in every case. Those hazardous substances that must be used should be recycled whenever possible. Care should be taken to ensure that the method of disposal for unrecyclable wastes is as harmless to the environment as possible.

Goals   In this investigation, you will **survey** your school to take inventory of the potentially hazardous wastes present on school grounds. You will then **research** ways to properly handle and dispose of those wastes and to reduce the amount of hazardous substances used by your school.

LAB WARMUP

Concepts   The Resource Conservation and Recovery Act (RCRA) of 1976 was enacted by Congress to regulate the transportation, storage, treatment, and disposal of hazardous wastes. A hazardous waste is any substance in any form (solid, liquid, or gas) that poses a hazard to human health or the environment if not properly handled and disposed of. RCRA considers a substance a hazardous waste if it meets any of the tests for ignitability, reactivity, corrosivity, or toxicity. *Ignitable* wastes are capable of catching fire if exposed to excessive heat or flame. *Reactive* wastes explode or react violently when exposed to water or air. *Corrosive* wastes dissolve other materials, such as containers. *Toxic* wastes are poisonous if ingested or inhaled.

Review   Section 19.2, Hazardous Wastes, and Section 25.1, The Global Ecosystem, should be completed before beginning this investigation. You should also understand the following terms before you perform this investigation.

hazardous waste   ignitable   reactive   corrosive   toxic   RCRA

Make a **prediction** about the outcome of this exercise and write it in the Lab Notebook.

MATERIALS (PER GROUP)

- reference books on hazardous wastes

PROCEDURE

1. Your group will be assigned an area of the school to survey. These areas may include the custodial closet, kitchen, offices, art room, science laboratories, copier room, and outside building maintenance. Record your assigned area in the Lab Notebook.

2. Survey the area for potentially hazardous substances. Look for warning labels such as "caution," "poison," "flammable," "caustic," "harmful or fatal if inhaled or swallowed," and "keep away from heat or flame." Try to determine what characteristic(s) the substance has that makes it potentially hazardous. **CAUTION: When surveying hazardous substances, do not touch any of the materials in the containers. If you are unsure as to the identity of the substance, notify your teacher.**

3. Record the name of the substance, whether it is ignitable (I), reactive (R), corrosive (C), or toxic (T), and the warnings on the container. Record your recommendations for handling and disposal of each substance. (Look on the label or refer to books on hazardous wastes.)

4. Add your survey results to the class data on the chalkboard. Study the other groups' results. As a class, write down a list of general recommendations regarding the handling and disposal of hazardous wastes for distribution to faculty, staff, and students.

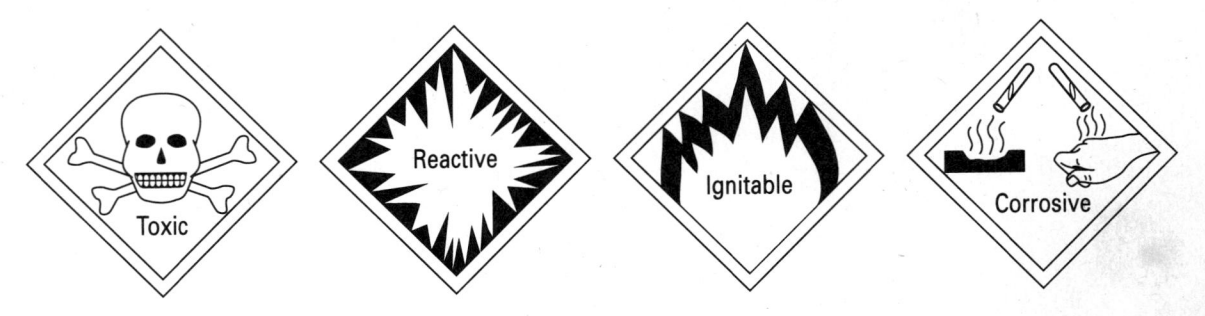

Figure 33.1 Typical warning signs for hazardous wastes

Name: _____ Class: _____ Date: _____

LAB NOTEBOOK: INVESTIGATION 33

PREDICTION A correct prediction is that hazardous substances will be found in all areas of the school. Steps can be taken to reduce the amount of hazardous substances used by the school, thereby limiting the need for disposal.

OBSERVATIONS

Assigned school location: _____

SURVEY OF POTENTIALLY HAZARDOUS WASTES

Potentially hazardous waste	Characteristic				Warnings	Recommendations
	I	R	C	T		
					Data will vary.	

DATA ANALYSIS

1. What characteristic of hazardous wastes was most common in your survey?
   Most of the substances surveyed were probably toxic.

2. What characteristic of hazardous wastes was least common?
   Answers will vary.

3. What area of the school contained the greatest amount of different hazardous substances?
   Answers will vary, depending on where the substances are stored.

4. What was the most common recommendation your class made regarding the hazardous wastes in your school?
   Accept all logical responses.

CONCLUSION

1. **Evaluate** How do the results of your survey compare with your expectations of the hazardous waste situation in your school?
   Answers will vary, but many students will be surprised at the amount of hazardous wastes in their school.

2. **Evaluate** Were hazardous wastes you surveyed adequately labeled? Explain.
   Answers will vary. Students might note that some hazardous wastes were not labeled or that some labels did not include directions for proper disposal.

3. **Integrate** What steps can be taken to make your school a safer, less hazardous place?
   Accept all logical responses. Students might suggest the use of safe alternatives to cleaning solutions and pesticides. A hazardous waste cleanup day could be declared to properly dispose of used hazardous wastes and their containers.

EXTENSION

Research Conduct a survey of hazardous wastes in your home. Encourage family members to reduce their use of hazardous substances, use safer alternatives, and dispose properly of hazardous wastes.

LABORATORY INVESTIGATION

34 ENVIRONMENTAL ISSUES AND PUBLIC POLICY

Problem: *Can an environmental issue be resolved to the satisfaction of all interested parties?*

INTRODUCTION

Background   It may appear that making policies to protect human health and the environment is a simple task. Industries could conserve resources or use only renewable resources. Unfortunately, an industrial society inevitably produces hazardous wastes, and their proper handling and disposal can be very expensive. Many resources are not recyclable and cannot be replaced by recyclable alternatives. Who pays for this? It may seem clear that these costs are the responsibility of the companies producing the wastes and using these resources. But when companies have to pay extra to deal with hazardous wastes, they must either raise the prices of their product or lay off employees to compensate for these costs and stay in business. Often, industrial plants provide widespread employment for local populations. Closing a plant due to environmental infractions can have devastating effects on the economy of the area.

Goals   In this investigation, you will **research** a hypothetical situation involving hazards to the environment and the residents of a community. You will then **model** one of the interested parties and report your findings to others, working with them to try to resolve the problem.

LAB WARMUP

Concepts   This scenario is set in a small, hypothetical midwestern city called Brockton. The area's major industry is a chemical company that employs half of the city's population. The plant moved to the city about a decade ago, and since the beginning of its operation it has been dumping its *effluent* (waste products) on the land near the Green River. Recently, toxic chemicals were found in the waters of the Green River and nearby Ruby Lake. Ruby Lake is the main source of water for the city. Its waters are also used for irrigation. These substances were traced back to the chemical plant's effluent.

An environmental agency of the state government has been made aware of the situation. This agency has demanded an immediate halt to the dumping of these hazardous wastes, since these chemicals are suspected of endangering the health of humans and wildlife in the area. The company argues that if it is forced to stop dumping, it will have to stop production at the plant and move somewhere else. Some of the city's residents want the dumping stopped; others fear that laying off half of the city's residents will place an unbearable economic burden on the city. The environmental agency insists on taking legal action against the chemical company. All interested parties are prepared to settle this dispute in court.

Review   Sections 25.1, The Global Ecosystem, and 25.2, Local Policies, should be completed before beginning this investigation. You should also understand the following terms before you perform this investigation.

effluent   public policy

Make a **prediction** about the outcome of this exercise and write it in the Lab Notebook.

MATERIALS (PER GROUP)

- library reference materials

PROCEDURE

1. Your teacher will ask your group to role-play one of the following interested parties: the chemical company, the environmental agency, the court, the residents who side with the chemical company, the residents who side with the environmental agency, and the city council. Record the interested party your group will represent in the Lab Notebook. Elect a spokesperson for your group.

2. Research the issue of dumping hazardous wastes, and acquire evidence to support your group's views. For example, if your assigned group is the court, research the laws regarding the dumping of hazardous wastes. Record your observations of the map of the area and your research notes in the Lab Notebook. Prepare your arguments for a town meeting and a possible day in court if the problem is not resolved.

3. At the next class meeting, have your group's spokesperson report your findings to the other groups. If no agreement is made among all interested parties, then the court will have to intervene and make a judgment.

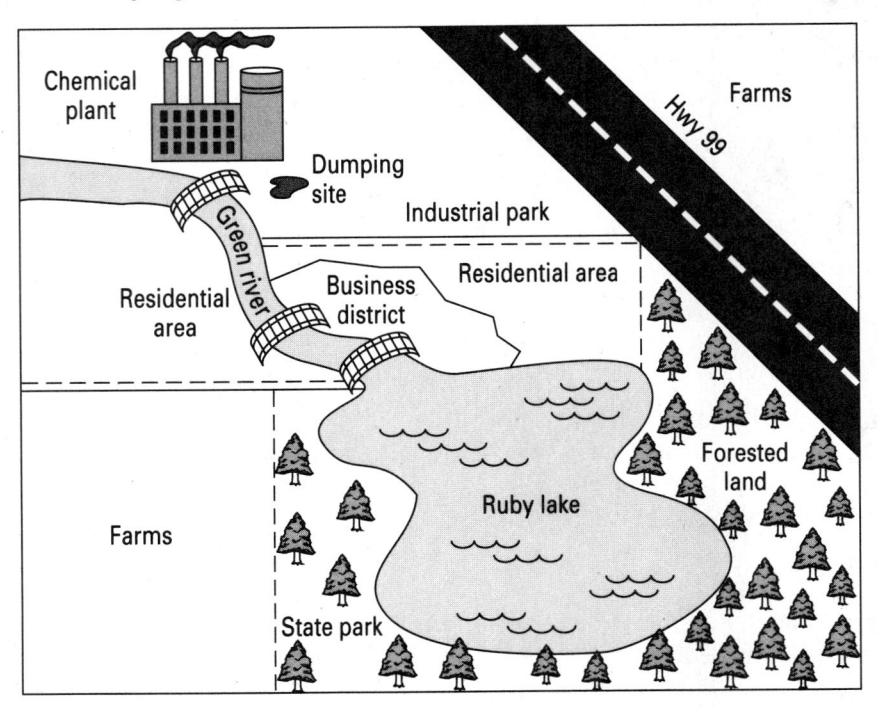

Figure 34.1 Map of area surrounding the city of Brockton.

Name: _____ Class: _____ Date: _____

LAB NOTEBOOK: INVESTIGATION 34

PREDICTION **A correct prediction is that environmental issues are not easily resolved, since there are many sides to any issue and many factors involved.**

OBSERVATIONS

Interested party: _____

Observations about the map: _____

Answers will vary. Students might observe that the Green River flows into Ruby Lake and that the city's residential areas are located along the river and the lake.

Research notes: _____

143

© Addison-Wesley Publishing Company, Inc. All Rights Reserved.

DATA ANALYSIS

1. What did your group propose to resolve the problem? What were your arguments?
   Answers will vary depending on student's assigned role.

2. What were some of the arguments of the opposing groups?
   Answers will vary.

CONCLUSION

1. **Integrate** On a separate sheet of paper, write an essay on the problems that arose when attempting to resolve this problem. Discuss factors involved in making a public policy that regulates industry. Refer to observations you made during this investigation to support your conclusions.

EXTENSION

Lobbying legislators through letter writing can be a very powerful way of changing public policy regarding environmental issues. Choose an environmental issue that interests you, research it carefully, and write a letter to your senator or representative offering your views about the policy. Make sure you investigate different arguments and understand the issue completely so you can support your opinions in the letter.